Introduction to
Grid Computing

CHAPMAN & HALL/CRC
Numerical Analysis and Scientific Computing

Aims and scope:
Scientific computing and numerical analysis provide invaluable tools for the sciences and engineering. This series aims to capture new developments and summarize state-of-the-art methods over the whole spectrum of these fields. It will include a broad range of textbooks, monographs and handbooks. Volumes in theory, including discretisation techniques, numerical algorithms, multiscale techniques, parallel and distributed algorithms, as well as applications of these methods in multi-disciplinary fields, are welcome. The inclusion of concrete real-world examples is highly encouraged. This series is meant to appeal to students and researchers in mathematics, engineering and computational science.

Proposals for the series should be submitted to one of the series editors above or directly to:
CRC Press, Taylor & Francis Group
4th, Floor, Albert House
1-4 Singer Street
London EC2A 4BQ
UK

Published Titles

A Concise Introduction to image Processing using C++
Meiqing Wang and Choi-Hong Lai

Grid Resource Management: Toward Virtual and Services Compliant Grid Computing
Frédéric Magoulès, Thi-Mai-Huong Nguyen, and Lei Yu

Introduction to Grid Computing
Frédéric Magoulès, Jie Pan, Kiat-An Tan, and Abhinit Kumar

Numerical Linear Approximation in C
Nabih N. Abdelmalek and William A. Malek

Parallel Algorithms
Henri Casanova, Arnaud Legrand, and Yves Robert

Parallel Iterative Algorithms: From Sequential to Grid Computing
Jacques M. Bahi, Sylvain Contassot-Vivier, and Raphael Couturier

Introduction to Grid Computing

Frédéric Magoulès
Jie Pan
Kiat-An Tan
Abhinit Kumar

CRC Press
Taylor & Francis Group
Boca Raton London New York

CRC Press is an imprint of the
Taylor & Francis Group, an **informa** business

A CHAPMAN & HALL BOOK

CRC Press
Taylor & Francis Group
6000 Broken Sound Parkway NW, Suite 300
Boca Raton, FL 33487-2742

First issued in paperback 2019

ISBN-13: 978-1-4200-7406-2 (hbk)
ISBN-13: 978-0-367-38582-8 (pbk)

Library of Congress Cataloging-in-Publication Data

Introduction to grid computing / Frédéric Magoulès ... [et al.].
    p. cm.
Includes bibliographical references and index.
ISBN 978-1-4200-7406-2 (hardcover : alk. paper) 1. Computational grids
(Computer systems) I. Magoulès, F. (Frédéric) II. Title.

QA76.9.C58I5772 2009
004'.36--dc22   2008049913

Warranty

Every effort has been made to make this book as complete and as accurate as possible, but no warranty of fitness is implied. The information is provided on an as-is basis. The authors, editor and publisher shall have neither liability nor responsibility to any person or entity with respect to any loss or damages arising from the information contained in this book or from the use of the code published in it.

Preface

With the first exploration in May 1999 by David Gedye in using a large number of Internet-connected computers as a supercomputer for searching extraterrestrial intelligence, the *Search for ExtraTerrestrial Intelligence (SETI)* project has been the first appearance of grid computing using a network of heterogeneous computers. This marks the beginning of grid computing, which is very different from parallel computing. This project is also the first in the subject of recovering unused computational cycles from computers in a network. Such a recovery of unused cycles has allowed *SETI@HOME* to gain access to 62 Teraflop/s in 2004. This is nearly double that of the most powerful computers in the world (36 Teraflop/s) in 2004. Through this, we have witnessed the immense computational power and great opportunity offered by grid computing. Recognized by the industry today, grid computing is gaining widespread adoption in various areas including customer relations, computational mechanics, biology and risk management in financial institutions. What we are seeing now is really a trend of increasing presence of grid computing comparable to that of electricity nowadays. As the technology matures, we believe that grid computing will follow the footsteps of the Internet to become more robust and accessible to the mass public in the near future.

This book aims at providing an introduction by illustrating state-of-the-art grid projects and technologies, and core grid technologies. This is wrapped up at the end by examples of potential applications of the grid.

In Chapter 1, we provide an introduction to grid computing and the concept of *virtual organizations (VOs)*. A comparison is made between grids and other distributed systems to bring out the advantages of grids and the motivations for using grids. The grid architecture is explained with respect to its main components to provide a background for subsequent chapters. We conclude the chapter with a discussion of some of the important standards used for implementing a grid.

The chapter, Grid Scheduling and Information Services, covers two important aspects of a grid system: scheduling of jobs and resource discovery, and monitoring grids. Scheduling is discussed with respect to both independent tasks (*metatasks*) and dependent tasks (*workflows*). For independent tasks, we describe some of the important mapping heuristics in the literature. This provides the background for understanding workflow scheduling.

A *static* workflow scheduling algorithm and an *adpative rescheduling* algorithm, which improves the performance of static algorithms, are explained. Scheduling algorithms, which consider the location of job-data while making scheduling decisions are also discussed. Fault tolerance strategies such as *rescheduling, job replication* and *pro-active* fault tolerance are discussed along with a framework that supports *workflow-level* and *task-level* fault tolerance. Grid workflow management systems, which form a layer between the user and grid middleware are discussed with detailed description of their components. Three workflow specification languages are explained with simple grid workflow examples. R-GMA and MDS are explained to give an idea about a grid information system. Components of an information system such as the data model, aggregate directory and service discovery are dealt individually.

Security in Grid Computing, Chapter 4, begins with a discussion of basic security concepts followed by a discussion of existing and emerging security technologies. In existing security technologies we discuss Public Key Infrastructure (PKI), which forms the basis of Grid Security Infrastructure (GSI) and Kerberos to explain the concept of network authentication protocols based on symmetric key. With respect to emerging security standards we discuss WS-Security and the OGSA security. We briefly explain WS-Security and the specifications that provide an extension to it. We also discuss how security issues pertaining to grids are addressed by OGSA. GSI describes the set of standards that provide security features to grid applications. Here we introduce the concept of *proxy certificates*, which are used for *single sign-on* and *credential delegation* in grids. An example of credential delegation over the network is demonstrated to illustrate how proxy certificates function in a grid.

Grid Middleware, Chapter 5, discusses the functions of grid middleware at the conceptual level. An overview of middleware together with the middleware services is presented to give a basic comprehension of the middleware concept. The notion of grid portals is also discussed to give readers a general idea of how heterogeneous resources on a grid can be used by end users with minimum knowledge on grids. The usage of these resources, from hardware such as telescopes to software such as databases, are made transparent through the use of grid portal. To illustrate these functions, several grid middleware applications adopted by practical scientific applications are presented. These middleware include: UNICORE, Legion, Condor, Nimrod-G, NINF/NINF-G, NetSolve, XtremWeb. Their presentations are organized according to functions and implementation methods. Some implementing skills used in grid construction and grid scientific applications, such as GridRPC, task farming, Peer-to-Peer, Portlet etc., are also specified.

Chapter 6 introduces in brief famous grid projects all over the world. Their research work is then described in detail. The research is classified into five technology aspects of grid computing: security, data management, monitoring, and information service collection and scheduling. This is in correspondence with the following chapters in the core technology section. While this chapter aims to provide a brief introduction, the following chapters offer a more detailed explanation of the mechanisms involved. Finally, the chapter concludes with a section on state-of-the-art applications of grid computing in present-day industry.

In Chapter 7, Monte Carlo methods, we first touch on some basic notions of Monte Carlo methods and the fundamental mechanism behind this method. Some groundwork on the generation of random numbers on the grid framework are also discussed and illustrated. According to experiment results, constraints on the model of random number generation require minimum communication between parallel computers, which is ideal for grid architecture. This model requires an additive Fibonacci matrix of which the dimensions have to be large to ensure the randomness of the numbers generated. This particular constraint requires that the model cannot be implemented on parallel structures as computation of the matrix on each node becomes too costly. Unless optimization methods can be applied to the matrix computation, the model for parallel generation of random numbers will not be feasible even on grid architectures. The experimental results on the computational time for sequential and parallel methods are compared to illustrate this imperfection. The parallel structure of the grid is then applied to industrial problems in the areas of finance and computational mechanics. Particularly in finance, we demonstrate the pricing of European options through the use of Monte Carlo method on grids (*gLite* and *Globus*). Besides providing an example on the actual implementation of the Monte Carlo method, these examples also allow users to perceive the differences between the two middleware *gLite* and *Globus*. While the scheduling of jobs to computer clusters is transparent to users in *gLite*, the allocation of jobs to computational resources in *Globus* requires the knowledge of Message Passing Interface (MPI) used to run parallel jobs on computer clusters.

In Chapter 8, Partial Differential Equations, different parallelization possibilities on the grid are demonstrated. They are namely data, time and spatial parallelization in the order of increasing complexity. While data parallelization induces zero communication overhead, time and spatial parallelization requires the communication of parallel computation nodes to evaluate the overall solution. The para-réel method is illustrated for time parallelization. This method gives us a first rough approximation of the solution followed by the refinement of the approximation towards the actual solution. This refinement is done using the parallel computation on the grid. While the para-réel method is used in time parallelization, the explicit finite difference method is employed in spatial parallelization. At each time iteration of the computation, the spatial domain is divided among the computational nodes for parallel computation. The computed results at each node are then reassembled to give an overall solution at a particular time step. This solution is then re-disseminated among the nodes to allow for the parallel computation of the solution at the next time step. These parallelization methods are then applied to the pricing of European options in finance. Similarly to the chapter on the Monte Carlo method, the actual implementation is demonstrated using C++ and MPI codes on *Globus*. While this implementation is specific to the heat equation, it is also applicable to the pricing of options as we also illustrate the transformation of the Black and Scholes equation to a heat equation. This transformation is used to allow for an easier implementation of the parallelization methods. Furthermore, such a transformation also allows for the use of radial basis functions and generalized fourier transform by leveraging on the symmetry offered by the heat equation.

Globus Toolkits is widely used software for building grids and implementing grid applications. Appendix A specifies this tool. Firstly, it gives a general description of Globus Toolkits 4.0 and its components. In this released version of GT, its components provide multiple functions, including resource monitoring and discovery, security infrastructure, job submission and data management. Some important and most frequently used components, such as Grid Security Interface (GSI), GridFTP, Reliable File Transfer (RFT), Replica Location Service (RLS), Data Replication Service (DRS), Grid Resource Allocation Management (GRAM), Monitoring and Discovery System (MDS) are discussed. The installation and configuration of GT4.0 are clearly specified in this appendix. To be more practical, we give a use case, where readers will understand how to define and submit the job, and how to monitor this job using GT4.0

The architecture and components of *gLite* are discussed in Appendix B to give readers a deeper understanding of this middleware. It highlights the importance of each component and the role it plays in the overall working of *gLite*. While the computing element (CE), storage element (SE) and workload manager service (WMS) are the main working components of *gLite*, other services such as the user interface and book-keeping services are also crucial to the functioning of the middleware. The basic usage of *gLite* is also discussed to provide readers with a first experience in *gLite*. Basic operations on the submission, collection and cancelation of jobs are demonstrated. The definition of a job description language (JDL) file to run sequential and parallel jobs is also illustrated so that readers can understand the JDL codes in the other chapters of the book.

Lastly, in Appendix C, we give a basic guide on the installation procedures of *gLite*. While the installation is monotone, and lengthy, the objective is to give readers a further understanding of the internal workings of each major component in *gLite*. For example, the option to install the computing element with or without the resource management system on the same cluster gives readers a better understanding of the role and internal composition of each *gLite* component. Moreover, readers will notice that during the installation of *gLite*, more components are illustrated than were discussed in Appendix B. This is due to the fact that each component discussed in Appendix B is made up of several other basic components in actual implementation. These additional basic components ensure the smooth functioning of each main component mentioned in Appendix B.

List of Tables

List of Figures

Contents

Chapter 1

Definition of Grid Computing

1.1 Introduction

Grid computing has emerged as an important field synonymous to high throughput computing (HTC) [1]. Contrary to other systems where the focus is to achieve greater performance measured in terms of the number of floating point operations the system can perform per minute, the importance of grids is defined in terms of the *amount of work* they are able to deliver over a period of time. The difference between high performance computing and high throughput computing has been illustrated by the *Condor* project [2]. Grids cannot be considered as a revolutionary technology. Rather they have evolved from existing technologies such as distributed computing, web services, the Internet, various cryptography technologies providing security features and virtualization technology. As we can see, none of these technologies is completely new. They have existed for quite some time and have been serving various needs. The grid technology takes features from these technologies to develop a system that can provide computational resources for some specific tasks. These tasks can be the simulation of stock markets to predict future trends, scientific research such as prediction of earthquakes or serving business needs for an organization having a geographically distributed presence. So in short, grid is an evolutionary technology, which leverages existing IT, infrastructure to provide high throughput computing.

One of the keywords that sums up the motivation behind evolution of the grid systems is 'virtualization'. Virtualization in grids refers to seamless integration of geographically distributed and heterogeneous systems. This enables users to make use of the services provided by the grid in a transparent way. This means that the users need not be aware of the location of computing resources. So, from the users' perspective, there is just one point of entry to the grid system. They just have to submit their service request at this node. Then it is up to the grid system to locate the available computing resources, which can serve the users' request. "Anatomy of the Grid" [3] introduces the concept of *virtual organization (VO)*. It defines a VO as a "dynamic collection of multiple organizations providing coordinated resource sharing". The formation of VO is aimed at utilizing computing resources for specific problem solving as discussed earlier. Based on the concept of VOs, we review three terms, which provide background for our understanding of grid systems. The first of these terms is *virtualization*, which has already been explained and stems from virtual organizations. The second term is *heterogeneity*. When we talk of VOs, it may imply that we are talking about a multi-institutional entity. The organizations that form part of a VO may have different resources in terms of hardware, operating system and network bandwidth. So, we infer that a VO is a collection of heterogeneous resources. The third term of importance is *dynamic*. Organizations can join or leave a VO per their requirements and convenience. So a VO is a dynamic entity. These three terms explain why grids have specific requirements as compared to other distributed systems. Ian Foster describes a three point checklist [4] to describe a grid. According to it, a grid should provide resource coordination minus centralized control, it should be based on open standards, and it should provide a nontrivial quality of service. A grid can be used for computational purposes (computational grid), for storage of data on a large scale (data grid), or a combination of both.

1.2   Grid versus Other Distributed Systems

In this section we bring out the major differences between grid and other distributed systems based on *Remote Method Invocation (RMI)* and Common Object Request Broker Architecture. Distributed systems generally serve the purpose of a single organization and have a centralized control. However, grids do not have centralized control and serve the purpose of a large number of organizations. A grid is defined by keywords such as heterogeneous resources, dynamic and virtualization (as explained in Section 1.1). Distributed systems may have heterogeneous resources but the extent of heterogeneity is limited to a single organization unlike grids, which are composed of heterogeneous resources from multiple organizations. A distributed system is static and has no concept of virtualization. Distributed systems focus on information sharing often using the client-server model. In grids the sharing is not limited to information. It may extend to applications and hardware. Distributed computing technologies enable information sharing within a single organization, whereas grids enable resource sharing among VOs (composed of multiple organizations). Grids support resource discovery and monitoring on a global scale. Such support is missing in distributed systems. If we consider decentralized systems like peer-to-peer systems, we observe that they provide very specialized services and are less concerned with quality of service. Further they do not have a notion of trust as in grid systems. Grids and peer-to-peer systems also differ on the basis of purpose, amount of data traffic and resources shared among the participating entities [5].

1.3 Motivations for Using a Grid

In this section we discuss the advantages gained by using grids over conventional systems. Some of these motivations stem from the definition of grid in terms of VO. The others can be explained in terms of the grid as a high throughput computing system. It is important to have an understanding of these concepts, as they form the basis for the architecture of grids.

1.3.1 Enabling Formation of Virtual Organizations

Grids enable collaboration among multiple organizations for sharing of resources. This collaboration is not limited to file exchange and implies direct access to computing resources [3]. Members of the grid can dynamically be organized into multiple virtual organizations. Each of these VOs may have different policies and administrative control. All the VOs are part of a large grid and can share resources. The resources shared among VOs may be data, special hardware, processing capability and information dissemination about other resources in the grid. As discussed in Section 1.1, VOs hide the complexity of the grid from the user, enabling virtualization of heterogeneous grid resources. Members of a grid can be part of multiple VOs at the same time. Grids can be used to define security policies for the members enabling prioritization of resources for different users.

1.3.2 Fault Tolerance and Reliability

Suppose a user submits his job for execution at a particular node in the grid. The job allocates appropriate resources based on availability and the scheduling policy of the grid. Now suppose that the node, which is executing the job crashes due to some reason. The grid makes provision for automatic resubmission of jobs to other available resources when a failure is detected. To illustrate this concept we take another example, data grids. A data grid can be defined as a grid for managing and sharing a large amount of distributed data. Data grids serve multiple purposes. They can be used to increase the file transfer speed. Several copies of data can be created in geographically distributed areas. If a user needs the data for any computational purpose, it can be accessed from the nearest machine hosting the data. They increase overall computational efficiency. Further, if some of the machines in the data grid are down, other machines can provide the necessary backup. If it is known in advance that a particular machine will be accessing the data more frequently than others, data can be hosted on a machine near to that machine. Both these examples illustrate the concept of virtualization. In the first example the user knows nothing about the grid failure. In the second example, the user accessing the data, does not know which machine in the system serves his/her request.

1.3.3   Balancing and Sharing Varied Resources

Balancing and sharing resources are an important aspect of grids, which provide the necessary resource management features. This aspect enables the grid to evenly distribute the tasks to the available resources. Suppose a system in the grid is over-loaded. The grid scheduling algorithm can reschedule some of the tasks to other systems that are idle or less loaded. In this way the grid scheduling algorithm transparently transfers the tasks to a less loaded system thereby making use of the under utilized resources.

1.3.4   Parallel Processing

Some tasks can be broken into multiple subtasks, each of which could be run on a different machine. Examples of such tasks can be mathematical modeling, image rendering or 3D animation. Such applications can be written to run as independent subtasks and then the results from each of these subtasks can be combined to produce the desired output. There are, however, constraints such as the type of tasks that can be partitioned in this way. Also there can be a limit on the number of subtasks into which a task can be divided, limiting the maximum achievable performance increase. If two or more of these subtasks are operating on the same set of data structures, then some locking mechanism similar to concurrency control in databases or semaphores in operating systems must exist so that the data structure does not become inconsistent. So there exists a constraint on the types of tasks, which can be made to run as a grid application and there also exists a limit to which an application can be made grid-enabled.

1.3.5   Quality of Service (QoS)

A grid can be used in a scenario where users submit their jobs and get the output, and then they are charged based on some metric like time taken to complete the task. In such scenarios where some form of accounting is kept for the services delivered to the user, a certain quality of service is expected by the user. This is specified in the *service level agreement (SLA)*. SLA specifies the minimum quality of service, availability, etc, expected by the user and the charges levied on those services. To be more specific, SLA can specify the minimum expected up-time for the system. As we have seen grids provide fault tolerance, reliability and parallel processing capability for certain tasks, and can be used to develop such distributed systems. Based on the requirement of the user, his/her task could be given priority over other users' tasks by the grid scheduling algorithm. For example, a user may require the services of the grid for a real-time application and thus has a more stringent QoS requirement than some other users. So, the grid scheduler could give his/her job more priority than other jobs and thus provide the necessary QoS to the user's real-time application. QoS can also be provided by reserving grid resources for certain jobs. If the resource reserved for a user's specific job is free for a while, it can report its status to a resource management node in the grid. The resource can then be used by the grid for its use until it is free. For example, if it is a computing resource, it may be used by the grid for execution of other jobs in the grid. As soon as the requirement for the reserved resource arises, the jobs utilizing these resources are preempted and make way for the higher priority jobs (the job for which the resources were reserved). The preempted job is put in the job queue along with the information on its completion status. This job can be scheduled by the grid scheduler once there are available resources in the grid. After reading this section, you might argue that there are other distributed systems that provide features like fault tolerance, sharing of resources, parallel processing etc. Then how is a grid different? Grids are different because they provide such features on a multi-institutional level and thus enable management of geographically distributed resources. Distributed systems that provide such features generally operate on an organizational level and have a centralized point of control unlike the grids.

1.4 Grid Architecture: Basic Concepts

Grid architecture refers to those aspects of a grid system that are taken into consideration when a grid is designed and implemented. Here we provide a brief introduction to these concepts to give the reader a foundation in grid concepts. These topics are covered in greater detail in subsequent chapters.

Grid architecture can be visualized as a layered architecture. The topmost layer consists of the grid applications and the APIs from a user's perspective. Then we have the middleware, which includes the software and packages used for grid implementation, for example *Globus Toolkit*, *gLite*. The third layer covers the resources available to the grid such as storage, processing capabilities and other application-specific hardware. Finally the fourth layer is the network, layer which deals with the network components like routers, switches, and the protocols used for communication between any two systems in the grid. In this section we discuss the components of middleware. They provide the basic functionality needed for grid computing.

1.4.1 Security

Just like any other system in the world, security forms the vital aspect of grid computing. We look at the three most desirable security features a grid should provide. These are single sign-on, authentication and authorization. Single sign-on means that the user is able to login once using his security credentials and can then access the service of the grid for a certain duration. Authentication refers to providing the necessary proof to establish one's identity. So, when you login to your email account, you authenticate to the server by providing your username and password. Authorization is the process that checks the privileges assigned to a user. For example, a website may have two kinds of user, a guest user and a registered user. A guest user may be allowed to perform basic tasks while the registered user may be allowed to perform a range of tasks based on his preferences. Authorization is performed after the identity of a user has been established through authentication. Other components of the grid that are part of security infrastructure are credential management and delegation of privileges. We discuss the grid components responsible for providing security feature in Chapter 4.

1.4.2 Resource Management

A grid must optimize the resources under its disposal to achieve maximum possible throughput. Resource management includes submission of a job remotely, checking its status while it is in progress and obtaining the output when it has finished execution. When a job is submitted, the available resources are discovered through a directory service (discussed in Section 1.4.4). Then, the resources are selected to run the individual job. This decision is made by another resource management component of the grid, namely, the grid scheduler. The scheduling decision can be based on a number of factors. For example, if an application consists of some jobs that need sequential execution because the result of one job is needed by another job, then the scheduler can schedule these jobs sequentially. The scheduling decision can also be based on the priority of the user's job as specified in the SLA (Section 1.3.5). We review resource management from a grid's perspective in Chapter 3.

1.4.3 Data Management

Data management in grids covers a wide variety of aspects needed for managing large amounts of data. This includes secure data access, replication and migration of data, management of metadata, indexing, data-aware scheduling, caching etc. We described replication of data in our discussion on fault tolerance. Data aware-scheduling means that scheduling decisions should take into account the location of data. For example, the grid scheduler can assign a job to a resource located close to data instead of transferring large amounts of data over the network, which can have significant performance overheads. Suppose the job has been scheduled to run on a system that does not have the data needed for the job. This data must be transferred to the system where the job will execute. So, a grid data management module must provide a secure and reliable way to transfer data within the grid. Grid data management is covered in Chapter 2.

1.4.4 Information Discovery and Monitoring

We mentioned that the grid scheduler needs to be aware of the available resources to allocate resources for carrying out a job. This information is obtained from an information discovery service running in the grid. The information discovery service contains a list of resources available for the disposal of the grid and their current status. When a grid scheduler queries the information service for the available resources, it can put constraints such as finding those resources that are relevant and best suited for a job. By relevance of resource we mean those resources which can be used for the job. If we talk about the computing capacity needed for a job and the job requires fast CPUs for its execution, we select only those machines fast enough for the timely completion of the job. The information discovery service can function in two ways. It can publish the status of available resources through a defined interface (web services) or it can be queried for the list of available resources. The information discovery service can be organized in a hierarchical fashion, where the lower information discovery services provide information to the one situated above it. The hierarchical structure brings about the flexibility needed for grids, which contains a vast amount of resources, because it can become practically impossible to store the information about all the available resources in one place. Grid information monitoring and discovery are discussed in Chapter 3.

1.5 Some Standards for Grid

In the previous section, we discussed the technologies needed in grid implementation. In this section we look at some of the open standards used for implementing a grid.

1.5.1 Web Services

As we shall see, grid services, defined by *OGSA*, is an extension of web services. So, grid service can leverage the available web services specifications. Here we discuss the most basic web service standards. The security-related web service specifications are discussed in Chapter 4. The four basic web service specifications are:

1. *eXtensible Markup Language (XML)* - XML is a markup language whose purpose is to facilitate sharing of data across different interfaces using a common format. It forms the basis of web services. All the messages exchanged in web services adhere to the XML document format.

2. *Simple Object Access Protocol (SOAP)* - SOAP [6] is a message-based communication protocol, which can be used by two parties communicating over the Internet. SOAP messages are based on XML and are hence platform independent. It forms the foundation of the web services protocol stack. SOAP messages are transmitted over HTTP. So unlike other technologies like RPC or CORBA, SOAP messages can traverse a firewall. SOAP messages are suitable when small messages are sent. When the size of message increases, the overhead associated with it also increases and hence the efficiency of the communication decreases.

3. *Web Service Definition Language (WSDL)* - WSDL [7] is an XML document used to describe the web service interface. A WSDL document describes a web service using the following major elements:

   (a) `portType` - The set of operations performed by the web service. Each operation is defined by a set of input and output messages.

   (b) `message` - It represents the messages used by the web service. It is an abstraction of the data being transmitted.

   (c) `types` - It refers to the data types defined to describe the message exchange.

   (d) `binding` - It specifies the communication protocol used by the web service.

   (e) `port` - It defines the binding address for the web service.

   (f) `service` - It is used for aggregating a set of related `ports`

4. *Universal Description, Discovery and Integration (UDDI)* - UDDI [8] is an XML-based registry used for finding a web service on the Internet. It is a specification that allows a business to publish information about it and its web services allowing other web services to locate this information. A UDDI registry is an XML-based service listing. Each listing contains the necessary information required to find and bind to a particular web service.

1.5.2 Open Grid Services Architecture (OGSA)

Open Grid Services Architecture (OGSA) defines a web services based framework for the implementation of a grid. It seeks to standardize service provided by a grid such as resource discovery, resource management, security, etc, through a standard web service interface. It also defines those features that are not necessarily needed for the implementation of a grid, but nevertheless are desirable. OGSA is based on existing web services specifications and adds features to web services to make it suitable for the grid environment. OGSA literature talks of grid services, an extension to the web services suitable for grid requirements. OGSA is discussed in Chapter 4, from a grid security perspective.

1.5.3   Open Grid Services Infrastructure (OGSI)

OGSA describes the features that are needed for the implementation of services provided by the grid, as web services. It, however, does not provide the details of the implementation. *Open Grid Services Infrastructure (OGSI)* [9] provides a formal and technical specification needed for the implementation of grid services. It provides a description of *Web Service Description Language (WSDL)*, which defines a grid service. OGSI also provides the mechanisms for creation, management and interaction among grid services.

1.5.4   Web Services Resource Framework (WSRF)

The motivation behind development of *WS-ResourceFramework* is to define a *"generic and open framework for modeling and accessing stateful resources using web services"* [10]. It defines conventions for state management enabling applications to discover and interact with stateful web services in a standard way. Standard web services do not have a notion of state. Grid-based applications need the notion of state because they often perform a series of requests where output from one operation may depend on the result of previous operations. WS-Resource Framework can be used to develop such stateful grid services. The format of message exchange in WSRF is defined by the WSDL. WSRF is supported by various companies and the specification has been finalized by the OASIS working committee.

1.5.5   OGSA-DAI

Open Grid Services Architecture-Data Access and Integration (OGSA-DAI) [11] is a project conceived by the UK Database Task Force. This project's aim is to develop middleware to provide access and integration to distributed data sources using a grid. This middleware provides support for various data sources such as relational and XML databases. These data sources can be queried, updated and transformed via OGSA-DAI web service. These web services can be deployed within a grid, thus making the data sources grid-enabled. The request to OGSA-DAI web service to access a data source is independent of the data source served by the web service. OGSA web services are compliant with *Web Services Inter-operability (WS-I)* and WSRF specifications, the two most important specifications for web services.

1.6   Quick Overview of Grid Projects

The research work of grid project is mainly for grid development. More and more engineers and scientists participate in this research field. They come from different discipline domains. Their work involves a great amount of scientific computations, which need a large quantity of computational resource and produce large scale data. As an example, in European Organization for Nuclear Research, known as CERN, a new instrument, named Large Hadron Collider (LHC), for discovering new particles is under research. LHC was put into operation in 2008. There are considerable experimental data generated each day by LHC. The processing of these data and the computation concerned with it are both so huge that they can not be completed by any one supercomputer or dedicated machine. Given this reality, grid technology was chosen as the solution to this challenge. Because of LHC, several research projects have started, for example, the European DataGrid project, the Enabling Grids for E-sciencE (EGEE) project, the National Institute for Nuclear Physics (INIF) grid project of Italy, the Grid Particle Physics (GridPP) project of the UK. As mentioned earlier, the Europeans are mainly focusing on grid-based high-energy physics work. In the United States, grid ultrastructural technologies have received much attention. The famous Globus project released the software tool Globus Toolkit, which has been commonly used in grid exploration. The local scheduler Condor produced by the Condor project has made significant contributions to high-throughput computing. In Asia, the ChinaGrid project of China, the BioGrid project of Japan and the GARUDA project in India have also done much meaningful work in both grid tools and grid applications.

1.6.1   American Projects

Globus [133] mainly works on grid infrastructure technologies. The core of Globus Grid is the toolset Globus Toolkit (GT). The current version GT4 has been released. GT comprises a set of layered grid tools realizing the basic services for security, resource location, resource management, communication, etc. These components have been deployed on top of Globus Ubiquitous Supercomputing Testbed (GUSTO) across 17 sites. They efficiently support the application grid infrastructure. The combination of Globus Toolkit and web service brings the future of a standardized grid research product.

Open Science Grid (OSG) [127] is an American grid infrastructure for scientific research. It organized a mass of computing and storage resources, and made them into a uniform shared cyberinfrastructure. Its 50 sites spread over USA, Asia and South America. It has two grids: Integration Grid and Production Grid. Integration Grid faces scientific research for its testing application and service. Production Grid faces industry and provides users with stable processing and data storage resources. One of OSG's motivations is to develop new services and then put them into the production environment. The current release version of OSG includes the services of Computing Element (CE), Storage Element (SE), Visual Organization (VO), Membership Service and Service Catalogue.

TeraGrid [138] is an ensemble of common high-end computational resources in the United States. These resources include high-performance computers and data resources distributed over 7 sites. A tool Common TeraGrid Software Stack (CTSS) has been developed for using these resources. CTSS is installed in all of the computers, which guarantees the homogeneity of services and tools on different resources: Inca can check the software version information of a computer resource and the results can be safely used by a web interface. Account Management Information Exchange (AMIE) realizes an automatic management of accounts. With respect to security, gx-map can manage a CA (Certificate Authority) of users.

1.6.2 European Projects

BeInGrid [132] (Business Experiments in GRID) is a European Grid project. Its objective is to lead the academic use and research of grids into the business sectors. Eighteen commercial experiments are going to be launched in the BeInGrid project. In addition, BeInGrid planned to develop a toolset repository of grid service components to well support European business. This software will fully profit from existing grid components in order to avoid re-developing.

EGEE [121] (Enabling Grids for E-sciencE) is a project aiming to provide computer resources for academic research and industrial production. The EGEE Grid is a worldwide grid. Users of this grid system are not limited by their geographical location. EGEE offers not only a stable and robust grid resource (30000 CPU, 5 petabytes of storage space), but also training services for its users. The applications of this grid system can be various. At present, its applications are mainly in two fields: high energy physics (HEP) and biomedical. More commercial and widespread applications will be launched on EGEE Grid in the future.

Grid5000 (France) [122] is a national grid project of France. It is a grid platform for academic research; 5000 CPUs distributed over 9 sites in France.

Users can reserve the PCs when they want to carry out their experiments. They can also configure the machines by themselves. This grid platform provides the mechanism of reservation and configuration to the users. Moreover, Grid5000 has offered a wiki-like web site for the communication of users. Users can submit their reports of experiments on this web site.

D-Grid initiative [123] is a German grid platform founded for education and research in 2005. Despite the contribution of a high performance resource of the grid, D-Grid is devoted to processing and accessing great amounts of scientific data. On this platform, a mass of scientific data, coming from various fields, such as high-energy physics, astrophysics, medicine etc, are collected and shared.

DutchGrid [125] is an open grid platform for research in the Netherlands. It provides a computing resource for various kinds of research experiment deployments. With respect to the security, DutchGrid Certificate Authority service, developed by NIKHEF in Amsterdam, allows the user to access or share the computing resource in the Netherlands or Europe.

GridPP [126] (Grid for Particle Physics) is a British project for a particle physics grid. The motivation of this grid project is to offer tools and infrastructure so users can transparently use the resources without searching for the resource themselves. The users are the physicists working for the LHC (launched in 2007), who need efficient cooperation and deal with massive data generated by the LHC. In fact, GridPP is a part of the project EGEE, and it is the UK's contribution to the LCG.

INFN [141] (Italy's National Institute of Nuclear Physics) is a research project that aims at the implementation and widespread use of a large-scale grid platform. In addition, INFN does much collaboration in Europe and all over the world, including CERN's LCG. INFN developed several middleware applications for distributed tasks scheduler and monitor, grid resource (computing resource and storage resource) management, user information collection, DataGrid, and web-based tools.

CrossGrid[124] project is a grid system with the function of realtime response. It enables users to monitor and control the application during the execution progress, for example, by changing its configurations. Most of CrossGrid's applications need interaction in realtime, such as the distributed realtime simulation of environment, which involves the interaction of doctors. The main applications of CrossGrid are in medical treatment, floods, particle physics and meteo/pollution.

CERN is famous for its huge invention of the World Wide Web. The Large Hadron Collider (LHC), the largest scientific instrument in the world, is now operational in CERN. The huge quantity of data produced by LHC is an enormous challenge for computer scientists. This task cannot be accomplished by any single computer. Because of the need to treat, store, and statistically analyze the massive quantity of data, the LHC Computing Project (*LCG*) [131] is launched by using computing grid architecture because of its easier maintenance of distributed systems and lower possibility of global failure (data are transferred and saved in several sites). But this architecture brings also some challenges, such as the assurance of communication among sites, management of heterogeneous hardware and softwares, data security and its sharing information management.

GÉANT [140, 139] project was cooperated by 30 European countries. It was composed of 26 National Research and Education Networks (NRENs). Its purpose was to build a huge backbone network at gigabit speed. This network was geographically distributed, but globally interconnected. IP service with QoS was offered by GÉANT. Project GÉANT ended in June 2005. A new network *GÉANT2* is now under construction. Similarly, GÉANT2 aims to build a huge scale network and provide advanced communication services. In addition, GÉANT2 adds some new research plans, such as "closing the 'digital divide"' and "examining the future of research networking".

DataGrid [142] is a project funded by the European Union. It is aimed at building the next generation computing infrastructure, which provides intensive computation and analysis of shared large-scale databases, from hundreds of terabytes to petabytes, across widely distributed scientific communities. DataGrid is focused on the high energy physics applications of CERN. It addresses the decomposed storage and handling issues of massive data. Then, the research results will be extended to other application areas, such as biology, earth observation and so on. DataGrid relies upon emerging grid technologies that are expected to enable the deployment of a large-scale computational environment consisting of distributed collections of files, databases, computers, scientific instruments, and devices. The GT platform is the supporting software under DataGrid software. In the DataGrid project, the developing work is divided into 12 work packages dispatched to 5 working groups: testbed and infrastructure, scientific applications, DataGrid middleware, project management and dissemination. The specification of task division can be found in reference [142].

1.6.3 Asian Projects

CNGrid [135] (China National Grid) is an important project supported by China. It is a testbed that integrates high-performance computing and the transaction processing capability of an information infrastructure. It effectively supports scientific research. CNGrid has developed grid-oriented supercomputers, and installed them in eight sites across the country. Ten subprojects of CNGrid cover different research fields, in which Scientific Data Grid (SDG) is included.

SDG [134](Scientific Data Grid) is based on mass scientific data resources. This project is aimed at connecting mass data resources of scientific databases, and sharing these geographically distributed, heterogeneous and autonomous data resources by means of grid technology. Some grid middleware for data access, information service, and security issues were used. These data involve the fields of astronomy, high energy physics and medical science.

ChinaGrid [119], also called China Education and Scientific Research Grid Project, aims to construct a public service platform for research and higher education in China. It is sponsored by 12 top universities, and established over the China Education and Research Network (CERNET). ChinaGrid Support Platform (CGSP) is the grid middleware developed for ChinaGrid. CGSP has implemented some complementary components that are not realized by Globus Toolkit.

NAREGI [137] (National Research Grid Initiative) is a Japanese cooperative project among industry, education, and the government, which is aiming to develop grid middleware and network technologies, including resource management, grid programming models, grid deployment tools, integration of grid software, network communication infrastructure, etc. In the field of industry, an application of nano-science technology is a portion of the project, with the objective to prove that the high-end grid computing environment can be utilized in nano-science.

BioGrid [136] aims to construct a datagrid, which not only gathers and processes massive databases and datasets, but also combines diverse computational resources into the data processes. It is initially designed for biological research in Japan. Its three main goals are deployment of an analyzer on the supercomputer network, seamless junction among databases and data processing, and data grid technology for linkages and operations among heterogeneous database systems.

GARUDA [129] is a cooperative project between science researchers and experimenters in India. Its objectives are to create a grid computing testbed and integrate the potential research and draw a more long term grid computing plan. The project's activities include construction of network, middleware, tools for managing computational resource and data, and web portal.

References

[1] Condor. High throughput computing. Web Published, 2007. Available online at: http://www.cs.wisc.edu/condor/htc.html (accessed January 1st, 2009).

[2] Condor. The Condor project. Web Published, 2007. Available online at: http://www.cs.wisc.edu/condor/ (accessed January 1st, 2009).

[3] Ian Foster, Carl Kesselman, and Steven Tuecke. *The anatomy of the grid: enabling scalable virtual organizations*, volume 2150 of *Lecture Notes in Computer Science*, pages 200–222. Springer, 2001. Available online at: http://www.globus.org/alliance/publications/papers/anatomy.pdf (accessed January 1st, 2009).

[4] Ian Foster. What is the grid ? A three point checklist. Web Published, 2007. Available online at: http://www-fp.mcs.anl.gov/~foster/Articles/WhatIsTheGrid.pdf (accessed January 1st, 2009).

[5] Ian Foster and Adriana Iamnitchi. *On death, taxes, and the convergence of peer-to-peer and grid computing*, volume 2735 of *Lecture Notes in Computer Science*, pages 118–128. Springer Berlin / Heidelberg, October 2003. Available online at: http://www.springerlink.com/index/CTQ1AGJECH3D55M3.pdf (accessed January 1st, 2009).

[6] SOAP. SOAP v1.2. Technical report, World Wide Web Consortium, April 2007. Available online at: http://www.w3.org/TR/soap/ (accessed January 1st, 2009).

[7] Erik Christensen, Francisco Curbera, Greg Meredith, and Sanjiva Weerawarana. Web Services Description Language (WSDL) v1.1. Technical report, World Wide Web Consortium, March 2001. Available online at: http://www.w3.org/TR/wsdl (accessed January 1st, 2009).

[8] OASIS. UDDI specification. Technical report, OASIS, 2007. Available online at: http://www.uddi.org/specification.html (accessed January 1st, 2009).

[9] S. Tuecke, K. Czajkowski, I. Foster, J. Frey, S. Graham, C. Kesselman, T. Maquire, T. Sandholm, D. Snelling, and P. Vanderbilt. Open Grid Services Infrastructure (OGSI) v1.0. Technical report, Global Grid Forum, 2007. Available online at: http://www.globus.org/toolkit/draft-ggf-ogsi-gridservice-33_2003-06-27.pdf (accessed January 1st, 2009).

[10] OASIS. Web Services Resource Framework (WSRF). Technical report, OASIS, 2007. Available online at: `http://www.oasis-open.org/committees/tc_home.php?wg_abbrev=wsrf` (accessed January 1st, 2009).

[11] OGSA. What is OGSA-DAI ? Web Published, 2007. Available online at: `http://www.ogsadai.org.uk/about/ogsa-dai/` (accessed January 1st, 2009).

Chapter 2

Data Management

2.1 Introduction

As grids are divided into computational grids and data grids, data management in grid environments should have double significance. In a comprehensive sense, it is applied to the computational grid. More narrowly defined it means the distributed data management system, which supports data access, synchronization and coordination of the distributed data across remote sites.

In the first definition, because there is less data transmission and it can be resolved by using small data files, data management is viewed as a less important problem because sometimes the data used by computing applications can be divided into small data files that make the scale of the data issue much smaller than the calculation issue. A frequently used solution is sending input data along with the executable file to the node where the calculation will occur. In the second definition, the data grid focuses on the processing of large amounts of distributed data. A typical example of a data grid application is data-intensive computation, which involves the processes of massive data storage, rapid data analysis and so on. Suppose that a traditional database server is adopted to perform a data-intensive application, and that a large quantity of data is produced in this procedure. In such a case, the database server comes to be a bottleneck because of its limited processing capability. One solution to this problem is applying a data grid, which distributes the generated data to dispersed sites (local or remote) and utilizes the capacity of individual resources to achieve a balance of work load. One of the most famous data grid research projects is DataGrid [12] launched by CERN, a European particle physics research institute, which has the objective of processing massive data produced by the Large Hadron Collider (LHC).

As a distributed database management system (DDBMS) and data grid are used under similar environments (physically distributed network), someone may mistakenly believe that they are the same thing. In fact, there are some differences between them. Firstly, data grid is completely heterogeneous, but this point is not explicitly put forward in distributed database systems. Heterogeneity, such as different data representation and different way, of data storage, is an important problem faced by data grids. In contrast, DDBMS has usually homogeneous data resources. Secondly, DDBMS can totaly control the data, but a data grid can only partially control data. For instance, operations used in DDBMS as, `insert`, `delete`, `update`, are all atomic operations. Atomic operations assure the consistency of all the concerned data. However, data grid cannot get full control of data resources. In grid environment, a data may be read by a user and at the same time be written by another user. Thirdly, the data resources of a data grid are much larger than the DDBMS's data resources, and a data grid should consider the scalability of data resources, which means that it should be feasible to add a new data resource.

2.2   Data Management Requirements

In the special environment of grids, data are geographically dispersed and heterogeneous in nature. The traditional data managing methods, for example, the insert, delete and update operations used in relational database management systems won't be appropriate. In the following section, we will first describe actual data characteristics in grids and then discuss the problems that should be addressed, that is the requirement of data management in grid environments.

2.2.1   Static Data and Dynamic Data

Data grids deal with two types of data. The first one is static data, which means that once these data are generated, they will only be read or analyzed, but never be modified or updated. An example of a static data grid is the DNA information that comes from original experiments, stored in one or more databases and will be only retrieved or compared with each other by the scientists. The other type is dynamic data, which involves dynamic updates and modifications. The data in enterprize-level e-business applications belong to this type. In a business data processing flow, every step has the possibility of changing the existing data. The potential operations include update, transaction of data operation, integration with external systems and synchronization.

In the case of static data, data operations are relatively simple. The common processing on these data are how to access the required data, how to move the required data to a certain node where the calculation needs them and how to effectively transfer the data. Along with the increasing complexity of the calculation on the grid, the concerned data changes from static to dynamic. The grid applications not only read the data but also write. A transaction of data operations performed on the data across multiple storage resource sites is also a common operation. The data is changing all the time. And under the grid environment, data is stored everywhere. One set of data may have several replicas at different sites. The synchronization among these replicas, in order for all replicas to have the real-time data, is a problem to be considered. Because data are stored in heterogeneous systems, a unified access to these data resources is an important factor. Furthermore, when a calculation needs the data from more data sources that are dispersed, integration of data from various storage sites such as database servers, file system servers must be performed.

2.2.2 Data Management Addressing Problems

Analyzing the processing of static and dynamic data in the distributed environment of a grid, we have outlined some issues, which should be addressed by data management, such as, data transfer that focuses on rapid and efficient data movement, data synchronization between the original and copied versions of dynamic data, and data integration that is used when the calculation needs data from more than one storage resource. In addition, there are two other points we haven't explicitly put forward: they are unified access to data and data replication. Considering data that are different in format, representation, and stored in diverse file systems or database systems, applications need a consistent manner to access them, and then data access should be independent of the actual implementation of data resources. Data replication means copying the original data and storing these replicas (data copies) at the node or near the node where they are more frequently used in order to reduce the overhead of network communication. We can summarize the main problems, which should be solved in data management as the following:

- Data unified access

- Data replication

- Data synchronization

- Data integration

- Data transfer

2.3 Functionalities of Data Management

2.3.1 Data Replication Management

Data replication is introduced into data grids as a method for optimizing data access [17]. Data replicas can be considered as a cache of data. Identical copies (replica) of data are created and distributed to various storage resource sites. Users or applications can access the nearest replica instead of looking for the original data and transferring them to where they are needed; therefore, the time dependence on data access latency is reduced. The responsibilities of a data replication management service (RMS) are:

- Create a replica for an entire dataset or part of a dataset

- Manage data replicas such as add, delete and modify the replica files

- Register a new replica into RMS

- Catalog the registered replicas so that users can query and access them

- Select the optimal replica according to the requirements of users or applications to best adapt their execution

- Assure consistency among replicas coming from the same dataset, automatically updating replicas once the original dataset is changed.

2.3.2 Metadata Management

Metadata is the descriptive information about the data. Metadata records information such as provenance information about how a data item is created or transformed, by which scientific instrument, physical information about their size, location, access authority and owners. There exist various metadata but they all include three main aspects of information as follows[15]:

- System information, which records the structural information about the data grid itself, such as service condition about the Internet, storage capacity of storage devices, computer idle status condition and usage policy.

- Replica information, which records the mapping relationship between a logical file and its physical copies.

- Application information, which records the data attributes that are specifically defined by one application community, for example, data content and structure, semantic information about the data item, and the circumstances under which the data were obtained.

Metadata is very important for retrieving, locating, accessing and managing the needed data in grid environments. Metadata management offers the ability to store and access the descriptive data and return to the user the desired attribute information about data items.

2.3.3 Publication and Discovery

The publication and discovery function is based on metadata service. In [16], they are described as two principal roles of metadata service. Data publication is a process of accelerating data research through precedent published information and making their associated attributes accessible to the user. Sometimes the publication process needs the metadata service to combine certain metadata information with the dataset so that they can be utilized in the data discovery service. Data discovery is the process of dynamically identifying a pertinent dataset and its location through querying attribute information published by metadata services, or other information such as the specification of internal structure, ownership and provenance or the physical properties (size, access path etc.) without specifying the identifier of the file or data item. After discovering the required data item, the data discovery service enables the user to further access the original data record or its replica. In addition, some discovery services are capable of displaying or visualizing the data resource contents as a schema to give the user a statistical view about the data.

2.3.4 Data Transport

Because of the data distribution in grid environments, the operations in data grids involve many data movements; therefore rapid and reliable data movement is an important aspect of data management. File transfer protocol (FTP) is one of the solutions; it realizes an effective transmission across networks, within an LAN or traversing WANs. In grids, a similar protocol, GridFTP, is adopted, which combines the Grid Security Interface (GSI) and provides secure, efficient data movement in grid environments. Actually, GridFTP is a successful prototype of grid data transfer. Based on its functions, we present its concrete aspects below:

- High speed transfer – supports various transport protocols over the network. For the purpose of high speed, data can be transferred in parallel.

- Stripped data transfer – large data files are partitioned into small blocks, and each data block is sent from one source node to a destination node. These blocks are then aggregated into the required data.

- Partial file transfer – transfer only part of a data file instead of the entire file.

- Third-party control of transfer – user or application can control a data transfer between two storage nodes from a third node. They can start a transfer and monitor and manage it.

- Restartable transfer – if data transmission is accidentally broken, it can be recovered from the point of failure and transferred sequentially.

2.3.5   Data Translation and Transformation

Data translation means changing the format of data with a little modification of information content. Data transformation means the derivation of the information in a different form (e.g. fourier transform). Data translation/-transformation is one function, which is necessary for the virtualization and uniform access of grid data. Data are stored on distributed sites, and lack standard for structuring, formatting, and representing. However, an application, which wants to perform uniform operations on these heterogeneous data, needs the data to be consistent. Another case is that the existing data format or representation does not match the needs required by an application or service. Most data resource accessed through a grid data service will need to undergo some form of translation and/or transformation to bring the data into an application useable state. Although application itself can also realize the function of translation/transformation by appending some special programming code, a best solution is to integrate this function into data management service.

2.3.6   Transaction Processing

Data transactions come from the database management system (DBMS). In DBMS, a calculation involving multiple data tables often utilizes transactions in order to protect data integrity. Transactions usually include numerous operations such as insert, delete, update, lock, and unlock. During execution of a transaction, once an error occurs or the user cancels the execution, which prevents the rest of the operations from sequentially proceeding, then all of the operations will be rolled back. This mechanism is implemented via a "two-phase commit" technique. Transaction processing has been proven very useful for data integrity and data recovery. In grid applications, the operations are not composed of simple database operations, but business processes that span several regions of control and multi-site collaborative operations. The interacting procedure is also more relaxed than in a traditional DBMS; operations occur in a more loosely coordinated manner. Grid transactions depend on asynchronous messaging instead of chained operations that occur one after another in a traditional DBMS [18]. Because of the huge amount of data that a grid data operation operates, the duration of transaction execution is considerably long, during which the possibility of failure is enhanced; thus fault tolerance and user control capabilities should be implemented.

2.3.7   Data Synchronization

In a static environment, replicas are all read-only; files are simply copied from one site to another. This method works well in static environments, and synchronization problems don't exist. However, in most grid environments, data is not only copied and transferred but also modified by users or applications. In the most ideal case, every distributed replica (either local or remote), including the original data, is kept totally consistent with each other, but this 100 percent consistency is impractical for a grid environment. In fact, sometimes users may not need this kind of consistency. We can relax the rule of consistency depending on the user's actual requirements and allow some part of the data to be inconsistent at some particular moments. Consequently, it is needed to define well the different levels of consistency. Such work has been done in [19], which describes five levels of consistency from the loosest to the strictest:

- **Possibly Inconsistent Copy** (Consistency Level -1) This is the lowest-level consistency and is the easiest to achieve; file replica is made during the execution of multiple write operations to this file. The resulting file doesn't match a state of the original file at any point in time.

- **Consistent File Copy** (Consistency Level 0) At this level, the content of a file can match a snapshot of the original file at a particular point in time. The following synchronization operation can satisfy this consistency level: a read shared lock should be added on the file being copied. This shared lock allows multiple read operations but only one write operation on the same file.

- **Consistent Transactional Copy** (Consistency Level 1) The replica is produced at a point when no write transactions were ongoing. This consistency level guarantees that there is no consistency problem within a single file, but it cannot be assured that when more than one file are concerned there exists no inconsistency among them. Because in the latter case one file may contain some pointers to an object included in an external file, if this external file object is removed by its local operation during the procedure of replication, then a new inconsistency will occur. The proposed synchronization operations are to produce all relevant replicas in the same transaction or delete the references spanning multiple files before replication.

- **Consistent Set of Transactional Copies** (Consistency Level 2) From this level, the consistency between multiple replicas is solved. If the replicas of interrelated files are produced in the same transaction, then there exists no inconsistency between multiple replicas at one site. But at this level, we cannot say a replica of the original file on another site is also up-to-date, because this replica is being managed by a different site. Some other operations may have occurred on this replica without sending notice to the original file. For the first site where the original file is stored, the status of a replica at another site is unknown or unpredictable. Therefore, the consistency between the two sites is not achieved at this level.

- **Consistent Set of Up-to-Date Transactional Copies** (Consistency Level 3) This is the strictest level of consistency. Each replica in the grid is kept identical to one another. Any write/read lock should be negotiated with all other sites under the same grid environment. Such a level of consistency is difficult to achieve. It requires all operational requests to be submitted through one single interface and non-grid access is not allowed.

2.3.8 Authentication, Access Control, and Accounting

In a computational grid, computing resources are protected in order to avoid malicious utilization or security problems. Similarly, in some communities grid data resources need to be guarded and only selected users have the right to access them. Security authentication and authorization solutions based on Grid Security Infrastructure (GSI) have well filled this responsibility. GSI entitles user rights by signing a credential of the user or application and then allows credential delegation for access to individual resources. An access control level problem should be considered when we talk about access to grid data. For instance, data access can be an operation of updating or deleting one record from some data table (low-level) or executing a multi-site transaction (high-level). In the first case, the user needs rights only over one data table, while in the second case the user should have access rights over all sites including, the data in question. Hence, data management should support a multiple granularity access control mechanism. Control decisions usually take into account both global and local rules and policies. In addition to the administrator of a grid system, the provider of a resource can also decide which user has which right. An accounting mechanism is used to record the history information of resource utilization, such as use duration, frequency of use, number of databases used, and so on. This information is collected to calculate the resource utilization ratio and also used to estimate future resource usage. This information is also helpful to make a decision about where to place the data replica.

2.3.9 Data Access and Storage Management

Data access and storage management are the basic functions of grid data management. The methods of data access are rather diverse. A distributed file system (DFS) organizes distributed files into a visual tree; files are branches or subbranches of the tree. Users can access files as if they are stored locally. Structural query language (SQL) is commonly used in traditional database systems, but it is not appropriate for data grids. However, some SQL-based methods are adapted to data grids. SQL statements are written into Java or C++ programs and compiled into executables to access data. SQL99 is used to access object-relational data and XML-based query works conveniently when a data resource is wrapped by a web service. Metadata services and replica services provide the useful information for accessing data files stored in the storage system. Furthermore, a uniform interface (or a set of API) to data resources is what users really need. A user inputs some parameters or query conditions and then gets the desired data returned. Storage management is responsible for creating a storage space on the system and writing data to it. The storage system may be a file system or database system; a dataset can be recorded to any storage system as well as storage media/device. The Storage Resource Broker (SRB) shows how diverse storage systems can be integrated and accessed under a uniform metadata-driven access mechanism. SRB can deal with requests by mapping to multiple storage systems of different types.

2.3.10   Data Integration

Data integration aims to collect and combine data from different resources and provides the user with a uniform view of these data [20]. Although here we view data integration as a grid data management function, in fact, it is a topic to be thought of as an independent technique. In principle, data integration consists of data discovery, data access, data transport, data analysis and data synthesis. Data discovery is the first step, and it asks the metadata service to find the relevant data. The next step is data access, which attempts to verify availability of these data and to see whether these data are useful for the problem to be resolved. Useful data will be read and transferred to the local host to be processed in the following step. Data analysis is an important step of data integration. It decides whether to do a combination between the data found in previous steps and local data or remote data. The potential analysis may exist at statistical and semantic levels. As each data source usually uses its own data representation, two different representations in two data sources may have one same semantic. As a consequence some research work focuses on how to resolve semantic conflicts between heterogeneous data sources. Data synthesis is a computing process that transforms a part of data and combines it with some existing data to produce a new view. This process give a new view of the old data so users can obtain new information.

2.4   Metadata Service in Grids

Metadata is data describing data and is not a new feature of grid computing. Here is an old example about metadata. Metadata can be used for managing book records in a library. In this example, metadata is the relevant information about a book and is written in a card, and these cards are collected so that the user can refer to them. Readers can find their desired book through looking up the information recorded on the card. Such information is something we are very familiar with: book title, key words, author name, and so on. In fact, such information is absolutely a kind of metadata. Metadata describes information about a data item itself and gives a richer signification so that data can be found more easily or be combined with other data according to a particular, similar attribute (metadata).

Why do we need such descriptive data (metadata) in a grid environment? The huge quantity of data stored in grids is the main reason. Firstly, traditional data research techniques (e.g., through SQL query or file name directly) cannot handle such a large amount of data. Secondly, data in grids are generally heterogeneous. They are different from each other in terms of storage format, data representation, and how they are controlled. Direct data discovery becomes much harder when faced with diverse data that cannot be accessed in a uniform manner. Thirdly, data from various scientific domains often have annotations by scientists. These annotations should be organized in an appropriate way in order to be associated with the original data as well as to be helpful for relevant data queries.

2.4.1   Metadata Types

As data are a very general concept, people have many ways to describe a piece of data. For a data file, we can describe physical characteristics such as file size, location in the storage system, and we can describe information about its replica(s) or a piece of the file content. It is necessary to classify metadata according to some categories. In fact, this classification is also the first job of all metadata software. References [22, 23] have illustrated their own classification for metadata, and our discussion about metadata types is based on their presentations. Relying on preceding works, we have summarized four types of metadata.

Data Metadata Data is a basic entity in a grid. Information about data is considered to be much more important than other metadata. Therefore, the taxology on it is very concrete in related work. We list four types of metadata about data.

- *Physical metadata* includes physical storage characteristics and properties about data including size of a data file or data object, location, creation time, creation owner, creation date and the format or file type (.doc, .txt, .jpg, etc.). Another metadata in this category is about natural data type. Each data management system has its own definition for data types, and unfortunately these definitions are not unique. As an example, in MySQL and Oracle, there exist two different names for the same data type (e.g., char and varchar); however, in the user's point of view, there is no difference in nature. A data type transformation metadata indicating the corresponding relationship between data types of diverse systems should be used, and include what data query could find all desired data on different data management systems across the whole grid.

- *Replica metadata* records link between logical data file and its one or more physical replica, which may be stored at different file systems or database systems. This type of metadata is often used by replica management systems (we will detail this point later). [1]

- *Domain metadata* describes a kind of data attribute used specifically in a domain that produces the data. In general, the terms or concepts used for this kind of metadata are coming from a set of conventions defined by scientists and engineers working in this domain. Domain metadata records the relationship between data items. The main relationship between data or data collection is the subsumption relation, which means a hierarchic classification for data. As an example, tea and coffee are two popular drinks, and tea can be classified as green and black. We can see from this example that green tea *is a* tea, and drink *is super class of* tea or coffee. Tea and coffee don't have subsumption relations between each other.

User Metadata The user is the one who creates, uses, or modifies the data and metadata. User metadata includes the user's information such as name, address, email, phone number, and password. Attribute *domain* specifies for which organization or project the user works, and *domain* can be just a name of the project or organization. Apart from *domain*, users can also be grouped by other criteria. One user can register himself into a *group*, and all users in one group have the same rights for resource access and resource control. Use of metadata facilitates the user rights management.

Application Metadata Data is generated by applications, and data is used as input and output of applications. Application metadata can represent content of data, the environment under which the data was produced, or some particular information for recording a data processing procedure. Sometimes, an application is broken into several application components, and the input data is also divided; therefore, some relationship information between these data used by the component applications should be constructed. Application metadata include the relationship information of data used by the whole application and its components.

Resource Metadata Resource is an important element used in data management. It is utilized in creating, storing and transferring data. Resource metadata describes characteristic information about the resource. It includes

---

[1] In reference [15], it is recommended to store replica metadata in the replica management system at the implementation level.

access address, physical location, resource type[2], and access control list (ACL) for resources.

2.4.2 Metadata Service

Metadata Storage There are two ways to store metadata; the one is a metadata database where metadata is stored in a relational database, and the other is storing metadata in an XML file. Using a database to store metadata, metadata can be accessed by standardized SQL statements. In grid environments, there is more and more metadata needed to be saved, so scalability is a problem. For this reason, a distributed metadata database system is more and more popular. In contrast, XML has the advantage of scalability in nature. XML can describe a data structure that is platform independent. It is free and extensible, which makes XML a flexible media for storing metadata. In addition, XML can be well combined with web technology, which facilitates the development of metadata systems.

Data publication and discovery Metadata service plays an important role in data publication and discovery. During data publication, data generated by diverse scientific instruments is calibrated and stored in a standard format; then it is made available to the community. Metadata services allow engineers or scientists to add metainformation, such as annotations, domain-specific metadata, to raw data, which facilitates data discovery. Data discovery is accomplished through posing a query to metadata service. In a query, the user can specify some characteristics about the desired file or dataset. Then this query is submitted to a metadata catalog, which will map from the query to the data with associated attributes.

2.5 Replication

Data replication originated in the concept of cache thought. A new replica is created because the new storage location provides better performance or accessibility for some particular application. A replica might be deleted because of storage space, lifetime, or other reasons. Data replication not only solves the bottleneck problem of grid data access, but also enhances the ability of efficient data access. In a grid environment with data application, data access is a cooperative precessing of diverse services. At first, the *metadata service* searches the required data according to the given attributes of desired data,

---

[2]Resource is classified into replicated-resources, striped-resources, write-once resources, read-only resources [22].

and it then returns a logical file name to the user or application. The logical file name is sent to the *replica management service*, which in turn sends back a list of locations of one or more replicas. The *replica selection service* will help to find the best replica location of a given dataset. It refers to the information provided by the information service and measures the performance of every location replica and estimates the transfer time of each replica point and finally selects and returns an optimal replica location to the user or application.

Replica metadata Replica metadata is also an important part of metadata in grid environments. Replica metadata includes information for mapping file instances to particular storage system locations [15]. As an example, in the project EDG[24], replica metadata stores mapping between the logical file name and Globally Unique IDentifier (GUID), a 128-bit number used to guarantee that data has a unique identifier in grid environments. Applications have to send their queries with some given metadata information to the metadata service in order to locate the physical file instance to access.

Replica catalog A replica catalog is frequently mentioned in grid data management. Replica catalog is a catalog service for registering and querying data replicas. Once a new replica is created, it will be registered into the replica catalog. If a replica is expired or there isn't enough storage space, it will be deleted, accordingly, the registry in the replica catalog will also be deleted. The replica catalog records the location information about the physical data instance. Once the user gives a logical file name, one or more physical data instances will be returned to the user.

Replica management Replica management is responsible for the creation and deletion of data replica at a storage site. Creating a replica might enhance the availability or accessing speed to data, while deleting a replica may be caused by the loss of disk space. In most cases, a replica is one exact copy of original data, which is created for better performance. Replica management involves the maintenance of the replica catalog.

Replica selection Replica selection is one of the current research axes of data replication. Replica selection means selecting an optimal replica from multiple replicas available. A so-called optimal replica is the replica that can offer high access speed, can economize the expenses of accessing and transferring data, or can get higher security. In short, the aim of replica selection is to obtain higher performance. Replica selection is a higher-level service, which is based on a set of basic infrastructure services such as, metadata service, storage system, and resource management [15]. Replica selection utilizes the mapping information offered by replica metadata and replica catalog for locating desired data instance. Apart from looking for a best replica from existing data instances, replica selection can also create a new data instance if the new location of this instance brings better performance. Another case of replica selection is subset selection. Most often, scientific experiments generate huge amounts of data; however applications need only a part of this data. Replica selection provides this type of application by creating an instance of the desired subset of data so as to reduce the data transfer burden.

Replica location Replica location refers to locating the storage position of a replica. Although the location replica catalog (LRC) records the location of every registered replica, users cannot obtain the physical location of a desired replica directly from the LRC because the LRC records the replica of only one local storage site. For locating a replica, a global view of all replicas is needed. In Giggle [25], a component named Replica Location Index (RLI) acts as a broker, which allows users to query replica location through it. RLI stores mappings between logical file names and the replica catalog, which exist in the form of an entry (LFN, pointer to an LRC). A, RLI may index one or more LRCs. One LRC can be indexed in one or more RLI. RLIs may be formed into hierarchical topology, that is one RLI could index the index of lower-level RLIs. Figure 2.1 shows the architecture of Replica location service formed by LRCs and hierarchical structured RLIs.

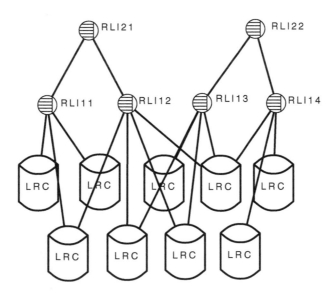

FIGURE 2.1: Replica location formed by multiple LRCs and RLIs in a two-level hierarchical structure.

2.6 Effective Data Transfer

In grid environments, data is stored on many distributed storage sites because a single storage device cannot hold a great amount of data generated by large-scale scientific experiments. These data might come from one experiment, but they may also come from several experiments, in which case it is possible that these data distributed geographically will be used in one calculation. It is common for users or applications to transfer data from one site to another. However, because data transfer doesn't create alternative data, users always want the data movement procedure to be as fast as possible. As we know, the network resource is limited and bandwidth is one of the biggest restrictions. In general, there are three types of bandwidth limitation [26]: the first limitation is the bandwidth of data provider server connected to the Internet, the second one is the bandwidth limitation between data provider server and client, and the third one is the bandwidth of the client connection to the Internet. In order to fully utilize limited network resources, especially its limited bandwidth, people turn to download/transfer data from multiple storage nodes in parallel. In the following sections, we illustrate several methods based on this principle.

Co-Allocation Data Transfer As the data replication technology is widely applied, one large data set has several copies (replicas), which are stored on several storage sites; thus transferring a dataset could be divided into multiple smaller transferring tasks. Each task transfers a part of the data from one replica location node to the destination node. Finally, these pieces of data transferred by multiple parallel tasks will be reassembled to compose a whole dataset at the destination end [26]. This is called a co-allocation data transfer.

Co-allocation transfer has several allocation mechanisms. There are two basic types of allocation mechanism, stateless allocation and stateful allocation. In stateless allocation, the client requiring the transfer must subscribe to a certain number of servers where the replicas are located. When the client transfers a dataset, this dataset is divided into several equal parts. Each part is transferred from one replica location site. Using this mechanism, it is not necessary to maintain the range of part and it is also convenient to extend it to dataset file with a size much larger; furthermore, it has good fault tolerance. However, the work for recomposing a dataset file is very complex, especially for large datasets. It utilizes all available replica location sites without any performance evaluation, and the bandwidth difference among various links are not exploited. This means the transfer allocation is not one that is well calculated, and the workload will not be balanced. In order to optimize this issue, those states of transfer condition and server performance should be considered.

Stateful allocation mechanisms consider the state of the replica location site, traffic condition of the network, different bandwidth, etc. It can decide dynamically how to allocate the data transfer on available replica sites according to these conditions mentioned above. In this branch, people have put forward several methods, such as historic-based allocation and dynamic co-allocation. In historic-based allocation mechanism, a prediction for transfer rate based on its history of previous transfer rate is a key point. The transfer rate is recorded as historical statistical data into a database. From these historic records, it is easy to predict the transfer rate of the next transfer. With this prediction, the size of the part which is partitioned on multiple servers can in turn be calculated. The faster a server is, part of the data file on it will be allocated. Each replica location server keeps sending data until the file is fully received. All servers will be kept busy during the period of transferring. To achieve maximum transfer speed, every server must finish its transmission at the same time, which is to say that the transfer speed should not change during a transmission. However, in practice, network conditions are often changed. If the condition of the network or bandwidth change rapidly, this method will not work as well as anticipated.

However, the dynamic co-allocation works much better than historic-based allocation when the transferring rate is variant, and it can well adapt to changing network conditions. In dynamic co-allocation mechanisms, the faster server transfers more data than the slower one. The required dataset file is partitioned into blocks that are of equal size. Each replica location server transfers one block at first; in this case, the faster server will finish transferring earlier, and then a new block with same size will be allocated to it. As a result, the faster server transfers more data than others. The maximum transfer speed is also achieved when all servers are kept busy until the whole dataset file is transferred. In this mechanism, there exists a round-trip-time, which means the period of time between when one block finishes transferring and the next block request is received. During this period, no data are transferred. To avoid this idle time, several block transfer requests to the same server could be made into a pipeline; thus once a block is transferred, another block request responds from the same server.

Peer-to-Peer Data Sharing Peer-to-peer emerges as a novel fashion of data sharing and downloading. In the peer-to-peer computing model, each node connected in a network can communicate with other nodes directly for sharing data and resources.

References

[12] DataGrid. Web Published. Available online at: `http://eu-datagrid.web.cern.ch/eu-datagrid/` (accessed January 1st, 2009).

[13] Michael Di Stefano. *Distributed data management for grid computing.* Wiley, 2005.

[14] William Allcock, Joseph Bester, John Bresnahan, A. Cervenak, L. Liming, and Steven Tuecke. Gridftp: Protocol extensions to ftp for the grid. Web Published, 2001. Available online at: `http://www-fp.mcs.anl.gov/dsl/GridFTP-Protocol-RFC-Draft.pdf` (accessed January 1st, 2009).

[15] Ann Chervenak, Ian Foster, Carl Kesselman, Charles Salisbury, and Steven Tuecke. The data grid: Towards an architecture for the distributed management and analysis of large scientific datasets. Web Published, 2001. Available online at: `http://loci.cs.utk.edu/dsi/netstore99/docs/papers/chervenak.pdf` (accessed January 1st, 2009).

[16] Gurmeet Singh, Shishir Bharathi, Ann Chervenak, Ewa Deelman, Carl Kesselman, Mary Manohar, Sonal Patil, and Laura Pearlman. A metadata catalog service for data intensive applications. Web Published, 2003. Available online at: `http://www.isi.edu/~deelman/mcs_sc03.pdf` (accessed January 1st, 2009).

[17] Leanne Guy, Peter Kunszt, Erwin Laure, Heinz Stockinger, and Kurt Stockinger. Replica management in data grids. Web Published, 2002. Available online at: `http://edg-wp2.web.cern.ch/edg-wp2/docs/ReplicaManager/ReptorPaper.pdf` (accessed January 1st, 2009).

[18] Malcolm P Atkinson, Vijay Dialani, Leanne Guy, Inderpal Narang, Norman W Paton, Dave Pearson, Tony Storey, and Paul Watson. Grid database access and integration: Requirements and functionalities. Web Published, 2003. Available online at: `http://www.ogf.org/documents/GFD.13.pdf` (accessed January 1st, 2009).

[19] Dirk Dllmann, Wolfgang Hoschek, Javier Jaen-Martinez, and Ben Segal. Models for replica synchronisation and consistency in a data grid. Web Published, 2001. Available online at: `http://www.cern.ch/hst/publications/replica_consistency.ps` (accessed January 1st, 2009).

[20] Maurizio Lenzerini. Data integration: A theoretical perspective. Web Published, 2002. Available online at: `http://www.dis.uniroma1.it/~lenzerin/DASI-School/materiale/dataIntegrationPapers/surveyPods2002.pdf` (accessed January 1st, 2009).

[21] Ian Foster and Robert L. Grossman. Data integration in a bandwidth-rich world. *Communications of the ACM*, 46(11):50–57, 2003.

[22] MCAT. MCAT-A meta information catalog (version 1.1). Web Published, 2004. Available online at: http://www.npaci.edu/DICE/SRB/mcat.html (accessed January 1st, 2009).

[23] Gurmeet Singh, Shishir Bharathi, Ann Chervenak, Ewa Deelman, Carl Kesselman, Mary Manohar, Sonal Patil, and Laura Pearlman. A meta-data catalog service for data intensive applications. Web Published, 2003. Available online at: http://www.isi.edu/~deelman/mcs_sc03.pdf (accessed January 1st, 2009).

[24] David Cameron, James Casey, Leanne Guy, Peter Kunszt, Sophie Lemaitre, Gavin McCance, Heinz Stockinger, Kurt Stockinger, Giuseppe Andronico, William Bell, Itzhak Ben-Akiva, Diana Bosio, Radovan Chytracek, Andrea Domenici, Flavia Donno, Wolfgang Hoschek, Erwin Laure, Levi Lucio, Paul Millar, Livio Salconi, Ben Segal, and Mika Silander. Replica management in the European datagrid project. Web Published, 2004. Available online at: http://hst.web.cern.ch/hst/publications/replica-management-grid-journal2004.pdf.

[25] Ann Chervenak, Ewa Deelman, Ian Foster, Leanne Guy, Wolfgang Hoschek, Adriana Iamnitchi, Carl Kesselman, Peter Kunszt, Matei Ripeanu, Bob Schwartzkopf, Heinz Stockinger, Kurt Stockinger, and Brian Tierney. Giggle: A framework for constructing scalable replica location services. Web Published, 2002. Available online at: http://www.isi.edu/~annc/papers/chervenakFinalSC2002.pdf.

[26] Sudharshan Vazhkudai. Enabling the co-allocation of grid data transfers. Web Published, 2003. Available online at: http://www.csm.ornl.gov/~vazhkuda/Vazhkudai-S-coalloc.pdf.

Chapter 3

Grid Scheduling and Information Services

3.1 Introduction

Scheduling and information services are two components of grids, which play an important role in the overall performance of an application running on the grid. The information services complement the grid scheduling system. They provide information about status and availability of resources in the grid. The resources in the grid can either be physical resources such as processors, memory and network bandwidth or a service offered by a node in the grid. Scheduling a job on a node requires two considerations. First, does the resource fulfill the minimum requirements and specific QoS requirements, if any, for the execution of the job? Second, is the resource available to serve the job? Both are provided by the grid information service. However, a scheduling decision is not as simple as that. We now present some cases that complicate the scheduling decision. A task may be composed of several sub-tasks, which are executed on different nodes. These sub-tasks may have dependency among themselves in terms of their order of execution. A scheduling algorithm must consider such dependencies while making a scheduling decision. As another example consider the scheduling of a job that has a very large input file. In this case the scheduling of a task to a node should not be made independent of the data location because significant communication overhead might be involved in transferring the data to the node executing the task. A node in the grid might fail due to a hardware failure or a network failure. In such cases the grid scheduler must reschedule the task onto a different node. Such a decision is made by consulting the grid information service. In this chapter, we cover the scheduling aspects for these examples as workflow scheduling, data-intensive service scheduling and fault tolerance.

The grid information service defines the format of the information stored in a directory. Because of the large scale of grids both in terms of geographical distances and the number of resources, a centralized architecture is not suited for the information service. It provides a single point of failure and is non-scalable. As we shall see in this chapter, a hierarchical structure of the grid information service helps it to overcome these limitations. The information service can be queried either directly by the user or by a scheduler to obtain information about the grid resources. So the design of a good execution schedule is dependent on the information provided by the grid information service.

3.2   Job Mapping and Scheduling

As discussed in the first chapter, a grid consists of a collection of heterogeneous resources. The objective of grids is the coordinated use of the heterogeneous resources to maximize the combined performance of these resources and to increase their cost-effectiveness. Because of the diverse nature of the resources, a grid can be termed a heterogeneous computing (HC) system. From here onwards we will be using the terms grid and HC systems interchangeably. Such HC systems can be used to solve computationally sophisticated problems, as discussed in [27]. The ability of HC systems to solve such problems can be attributed to their ability to match the computational resources to the computing needs [28]. In such diverse environments, not all machines are suitable for every task. Some tasks have specific machine requirements, for example, the need for a specific instruction set. So to minimize the overall execution time of the tasks and thus increase the throughput of the system, it is important that correct resources be assigned to every task. There are two ways in which tasks are assigned to HC systems. One way to exploit the complete potential of a HC system is to split the task into multiple subtasks. Each of these subtasks can then be assigned to a machine fulfilling its machine specific requirements. These tasks may have dependency among themselves. For example one of the tasks may need the data generated by another task. The first step after identifying the subtasks is to assign a machine to each subtask. This process is termed matching. After this the order of execution of the subtasks is identified. During this ordering, the intertask dependencies such as data transfer and inter-process communication are considered. This process of ordering the execution of subtasks is called scheduling. The overall process of matching and scheduling is termed mapping. The other way to do mapping is to map a collection of independent tasks. Such a collection of independent tasks is called a metatask. Note that individual tasks in the metatask may have data dependency among its subtasks. Each task in the metatask may have an associated deadline or priority [28]. An example of metatasks can be tasks submitted by different users for execution to the grid system.

The aim of mapping is to maximize an objective function, which is based on QoS attributes such as execution time, response time or those requested by the users of the HC system [28]. The general task of mapping jobs to machines is NP-Complete [29]. So certain heuristics have been developed for mapping. In the next section we talk about these mapping heuristics and the assumptions made while describing the heuristics. We also provide an idea about the relative performance difference among these mapping heuristics.

3.2.1 Mapping Heuristics

The heuristics for mapping the tasks can either be *static* or *dynamic*. Static mapping heuristics mean that the matching and scheduling decisions are made before the execution of the application. In dynamic mapping heuristics the matching and scheduling decisions are made during the application execution. Another classification of the mapping heuristics is based on the time when a task is mapped to the machine. If the task is mapped onto the machine as soon as it arrives, it is *on-line* mode. In *batch* mode mapping heuristics, the tasks are mapped at specific prescheduled times, termed *mapping events*. The set of tasks (metatask) mapped during a mapping event include tasks that arrived after the previous mapping event and tasks that were mapping during earlier mapping events but did not begin their execution [28].

Reference [28] defines mixed-machine HC systems as a collection of heterogeneous machines interconnected by high speed links. In such an environment the execution time of a machine can vary widely from machine to machine. Static mapping heuristics make an assumption that the estimated execution time of a task for each machine in the HC system is known in advance. For cases when two tasks may have a data-dependency among themselves, static heuristics assume that the data-transfer times are known in advance. The accuracy of static mapping heuristics depends on the accuracy of these estimates. Reference [30] talks about estimating the expected execution time using task profiling and analytical benchmarking. In task profiling, we analyze the type of computations present in an application by breaking the source program into homogeneous code blocks. This decomposition is based on the computational needs of the code blocks. The code types are defined based on the features of the machine architecture in the HC suite and applications being considered for execution by the HC system. Analytical benchmarking specifies the performance of machines in the HC suite on different code types.

The scheduling decisions in the grid systems are based on the mapping heuristics. A mapper can be defined as the component of the grid scheduler that runs the mapping algorithms. The mapper maintains a matrix known as the expected time to compute (ETC) matrix. It contains the expected execution times of a task on all the machines available in the grid. The entries in the row of an ETC matrix indicate the execution time of the tasks on different machines. The column entries specify the time taken by a machine to execute various tasks in the metatask. The variance in the execution times of the tasks in the metatask on a given machine is termed task heterogeneity. Similarly, the variance in the execution time of a task on different machines is termed machine heterogeneity. The ETC matrix can be divided into *consistent* and *inconsistent* types. For a consistent ETC matrix, if machine m_i has lower execution time than machine m_j for a task t_a, then the same is true for any task t_b in the metatask. For inconsistent ETC matrix, no such ordering can be made. However, a subset of an inconsistent ETC matrix may be consistent. Such ETC matrices may be termed *semiconsistent*. The mapping heuristics behave differently under different heterogeneity conditions. We will elaborate upon this as we discuss various mapping heuristics. We now explain some terms that will be used in our discussion of mapping heuristics.

- *Expected Execution Time:* This is the estimated time for the execution of a task on a machine when the machine has no load. e_{ij} denotes the expected execution time of task t_i on machine m_j.

- *Expected Completion Time:* It indicates the wall-clock time at which machine m_j completes the execution of task t_i. This includes the time needed to complete any previously assigned task to machine m_j before task t_i was assigned to it. It is denoted by c_{ij}.

- *Machine Availability Time:* It is the earliest time when a machine becomes free. This means that it has executed all the tasks previously assigned to it. It is denoted by mat(j). So $c_{ij} = mat(j) + e_{ij}$ can be derived from the above definitions.

In [28] the authors define makespan as $max(c_{ij})$. It denotes the maximum time taken for all the tasks in the metatask to complete execution. It is a measure of the throughput of the system. It, however, does not address the QoS requirements of the individual tasks. We now discuss the important mapping heuristic techniques described in the literature.

3.2.1.1 Opportunistic Load Balancing

Opportunistic load balancing simply assigns tasks to the next available machine. If more than one machine is available, one machine is chosen arbitrarily. It does not take into account the expected execution time of the task on that machine. As expected, this algorithm performs poorly for consistent, inconsistent and semi-consistent ETC matrices.

3.2.1.2 Fast Greedy or Minimum Completion Time (MCT)

This heuristic assigns each task in arbitrary order to the machine that has minimum completion time for that task. As this heuristic does not take into account the execution time of a task on a machine, some of the tasks may not be assigned to the machine that has the minimum execution time for those tasks. [31] shows that fast greedy heuristic performs well for inconsistent and semi-consistent ETC matrices in general but does not perform well for consistent ETC matrices.

3.2.1.3 User-Directed Assignment or Minimum Execution Time (MET)

In this approach each task is assigned in an arbitrary order to the machine that requires least execution time for that task's execution. As this heuristic does not consider machine availability time, it can cause severe load imbalance for consistent ETC matrices. However, compared to consistent ETC matrices, this approach fares better when the ETC matrix is inconsistent. Its only advantage over MCT is its simplicity of implementation.

3.2.1.4 Switching Algorithm

The switching algorithm (SA) heuristic [32] uses MCT and MET heuristics in a cyclic fashion in an attempt to balance load across machines. This heuristic tries to strike a balance between the two approaches. The SA heuristic is based on the following idea: The MET algorithm can create a load imbalance whereas the MCT heuristic tries to balance the load by assigning tasks to machine that provide earliest completion time. Reference [32] defines *load balance index*, π, as r_{min}/r_{max} where r_{min} and r_{max} stand for minimum and maximum machine availability time over all the machines. It can have a value in the interval [0,1]. Two threshold values for load balance index are chosen as π_l and π_h such that $\pi_l < \pi_h$ and both lie in the interval [0,1]. Initially π is set to zero. The SA heuristic begins using the MCT heuristic until the value of π equals π_h. Then the SA heuristic uses MET causing the load balance index to decrease. When it becomes π_l, the SA mapping heuristic switches back to MCT and this cycle continues. In this way the SA mapping heuristic uses the best of MCT and MET heuristics.

3.2.1.5 Min-Min

The min-min mapping heuristic [28] is based on the idea that by scheduling the tasks that change the machine availability time by the least amount it is likely to produce a small makespan for the metatask. The algorithm starts by computing the expected completion time for all the tasks over all the machines. It then computes the minimum completion time for each task in the metatask. The minimum of these completion times is chosen over all the tasks and that task is assigned to the corresponding machine. The machine availability time is updated by adding the expected execution time of the task to the current machine availability time. The assigned task is removed from the metatask list and the procedure is repeated until there are no tasks in the metatask list.

3.2.1.6   Max-Min

The max-min mapping heuristic [31] is similar to the min-min mapping heuristic. The first step of this heuristic is identical to the min-min heuristic. In the second step, instead of choosing the task having the minimum of earliest completion times among all the tasks, max-min chooses the task having maximum of earliest completion times and assigns it to the corresponding machine. The machine availability time is updated and the process is repeated for every task in the metatask. The max-min mapping heuristics would generally outperform min-min mapping heuristics, when the number of short tasks is greater than that of long tasks. As max-min chooses the longer task, it would first schedule all the long tasks. Then the shorter tasks could execute concurrently with the longer tasks. On the other hand the min-min heuristic would schedule all the smaller tasks first and then execute the longer tasks, thereby increasing the makespan of the metatask as compared to the max-min heuristic.

3.2.1.7   Genetic Algorithm (GA)

In [31] authors illustrate the use of a genetic algorithm for mapping tasks to machines in the HC environment. The algorithm starts with a set of chromosomes. Each chromosome is represented as a one-dimensional vector having a dimension equal to the number of tasks in the metatask. The *ith* entry represents the machine to which task t_i has been mapped. The initial chromosome population is generated in one of two ways: (a) randomly generated chromosome from a uniform distribution, or (b) one chromosome generated using a min-min mapping heuristic and the remaining are generated randomly. After the generation of the chromosome set, the population is evaluated based on its fitness value, which happens to be makespan. This means that the smaller the makespan of a chromosome, the better suited it is to be in the next generation of chromosomes. The genetic algorithm can be represented using the following pseudocode [33]:

```
initial population generation;
evaluation;
while(stopping criteria not met){
    selection;
    crossover;
    mutation;
    evaluation;
}
output the best solution;
```

The algorithm is stopped either after a pre-defined number of iterations or when the *quality* of chromosomes become almost static. Here quality refers to the makespan of the chromosomes. The selection of chromosomes is made on the basis of their rankings. The crossover selects a pair of chromosomes and chooses a random index from the first chromosome (represented by a vector). From that index onwards the machines assigned to the tasks for each of the chromosomes are swapped. The chromosomes are selected for crossover with a certain fixed probability. Again, for mutation, a chromosome is randomly selected and a task within the machine is randomly reassigned to another machine. At the end of the while loop, the new chromosome set is evaluated and the process repeats. The GA mapping heuristic performs best for all three ETC matrix types. If, however, the execution time of the mapping algorithm is the criteria for choosing the mapping heuristics and the ETC matrix is consistent, the min-min approach is preferred [31].

Other mapping heuristics include simulated annealing (SA), genetic simulated annealing (GSA), tabu search and A^*. Simulation results from [31] show that GA performs the best, followed by min-min and A^*. A^* does typically well for inconsistent matrices. Its performances, however, deteriorates for consistent ETC matrices. All these mapping heuristics assume that a correct estimate is available for the expected execution time. Reference [34] shows that even when correct estimates for execution time are not available *intelligent* algorithms like GA, min-min and min-max perform better than naive algorithms like opportunistic load balancing.

3.2.2   Scheduling Algorithms and Strategies

Scheduling of jobs on a single parallel computer differs significantly from job scheduling in a grid environment or metacomputer [55], which is defined as a collection of independent resources linked by an interconnected network. The task of scheduling on grids is complicated by the fact that many machines are involved, each having a different local policy. A metacomputer scheduler or grid metascheduler is implemented on top of local job schedulers. It is the responsibility of the metascheduler to schedule individual jobs to different local schedulers. The local scheduler may then schedule these jobs based on local scheduling policy.

A grid scheduling system can be divided into three parts: a scheduling policy, an objective function and a scheduling algorithm [52]. The *scheduling policy* is defined by the owner of the machine or the administrator of the organization owning the machine. It consists of a collection of rules to define resource allocation for the jobs submitted to that machine. For example, in an organization, a scheduling policy may give jobs from department A more priority than jobs from department B. So if jobs from both the departments are submitted at the same time, the job from department A would be scheduled ahead of the one from department B. *Objective function* assigns a numerical value to the schedule and helps in selection of a schedule from more than one possible schedule. Generally, an objective function consists of more than one parameter, which the scheduling system seeks to maximize or minimize. *Scheduling algorithms* form the heart of scheduling systems. A good scheduling algorithm should produce a near optimal schedule with respect to the chosen objective function and should not require too much resource or time for its execution. We start our discussion with the most basic scheduling algorithm, first-come-first-served.

3.2.2.1   First-Come-First-Served

First-Come-First-Served (FCFS) is a well-known scheduling scheme. In FCFS jobs are ordered based on their submission time. Then a greedy list scheduling approach is employed in which the next job in the list is started as soon as the resources are available for its execution. FCFS offers several advantages. It is a fair algorithm because the completion time of a job is independent of any job submitted after it. It does not require any knowledge of the execution time of the job on a particular resource. Lastly, FCFS is easy to implement and requires very little computational effort. The disadvantage of FCFS is that it may produce schedules that have a large percentage of idle time for the nodes. The performance of FCFS is further degraded if many jobs having large node requirements are submitted. Performance of FCFS can be improved by *backfilling*, which is the next topic of our discussion.

3.2.2.2   Backfilling Algorithm

Backfilling algorithm has been explained in [53]. This algorithm requires an estimate of job execution time on the machine and can be used with any greedy list schedule. If the job at the head of the list cannot be started due to the lack of available resources, then backfilling tries to find another job that can be started with the available resources provided that it does not delay the scheduled start of the job at the head of the list. Basically, backfilling tries to fill the holes in the processor's time slots. The important factors that affect the scheduling of a job are run-time of the job (length) and number of nodes required by the job (width). Two variants of backfilling are available: conservative and aggressive.

- *Conservative Backfilling:* In this variant of backfilling, resources are reserved for every job when it arrives at the queue. The jobs are allowed to move ahead in the queue as long as any queued job is not delayed beyond its reserved start time. It allows less backfilling than the aggressive approach due to constraints on the schedule of all the waiting jobs. Also, for longer jobs, it is difficult to get a reservation ahead of the previously reserved jobs. So, it is difficult to backfill longer jobs under the conservative strategy. However, wider jobs benefit from this approach as they are provided a reservation as soon as they enter the queue and thus guaranteed a start time for them.

- *Aggressive or EASY Backfilling:* In this backfilling, only the job at the head of the queue is given a reservation. Jobs are allowed to move ahead of reserved jobs provided they do not delay the scheduled start of that job. Longer jobs can backfill more easily than in conservative backfilling because of the presence of only one *blocking* reservation in the schedule. However, wider jobs suffer when this approach is used because they are not given a reservation until they reach the head of the queue. Low priority jobs can backfill ahead of them if they find enough free processors.

A variant of the backfilling algorithm has been proposed in [54], which tries to bring out the best among the two backfilling approaches. It groups jobs into categories and studies the effect of the two approaches on different job categories. It provides a reservation selectively to those jobs that have waited long enough in the queue. The amount of backfilling is greater than in conservative backfilling due to fewer reservations. By assuring reservations to jobs that have waited long enough, it mitigates the disadvantage of large delays caused by aggressive backfilling. To study the performance of both the backfilling approaches on jobs of different length and width, jobs are classified into four categories: Short Narrow (SN), Short Wide (SW), Long Narrow (LN) and Long Wide (LW). Jobs from the LN category benefit from aggressive backfilling and SW jobs benefit from conservative backfilling. For SN jobs there is no consistent trend among the two backfilling schemes as they can backfill very quickly in both the schemes. For LW category jobs there is no clear advantage of using one scheme over the other because aggressive backfilling allows them better backfilling opportunities and the conservative approach provides the advantage of reservations. So, the overall performance of both the backfilling algorithms depends on the composition of jobs with respect to the job categories. Based on these results, a selective reservation scheme is used, which uses two queues with different scheduling policies. A *'no-guarantee'* queue in which the start time of the job is not guaranteed and an *'all-guaranteed'* queue in which all the jobs are given a start time through reservation (similar to conservative backfilling approach). When a job enters a system it is put in the 'no-guarantee' queue. When the wait-time of the job exceeds a threshold it is transferred to the 'all-guaranteed' queue.

3.2.2.3 List Scheduling Algorithms

The basic idea behind list scheduling algorithms is assignment of priorities to the nodes of a *Directed Acyclic Graph (DAG)*. The nodes are then arranged in a list in descending order of their priority. A node having higher priority is examined for scheduling before a node having lower priority. If more than one node has the same priority, the tie is broken using some method like choosing a node arbitrarily. *Heterogeneous Earliest Finish Time (HEFT)* is an example of a list scheduling algorithm. Like all the list scheduling algorithms it consists of a node prioritization phase followed by a processor selection phase. A variant of HEFT used for adaptive rescheduling of workflows is discussed in Section 3.4.3.

3.2.2.4 Level-based Scheduling Algorithms

Level-based scheduling algorithms partition the DAG into levels of independent nodes. Within each level greedy, min-min, max-min or sufferage heuristics (Section 3.2.3) can be used to map nodes to processors. The schedule that produces the minimum makespan is chosen as the final schedule. Levelized Heuristic Based scheduling (LHBS) is a level-based algorithm for grids. The complexity of LHBS using only the greedy heuristic is $O(v * p)$ where v is the number of nodes and p is the number of edges in the DAG. The complexity of LHBS using the other three heuristics is $O(v^2 * p)$; see reference [56]. LHBS is discussed in more detail in Section 3.4.3.

3.2.3 Data-Intensive Service Scheduling

Applications like high-energy physics access and generate a large volume of data. The data transfer time required for such applications may be very large and may affect the overall execution time of the application. So it is important that scheduling decisions take into account the location of data for data-intensive applications. An important optimization in this regard would be the replication of data from its primary position to other repositories, reducing the data access time and frequency. Another way of doing this is to schedule the job to a processor near the data location. So, for data-intensive applications, the scheduling algorithms should not only consider the best mapping between jobs and idle processors but also proximity to data location. But neither of these approaches is perfect because they do not scale well in grids. Most of the scheduling algorithms have placed more importance on scheduling of jobs and have considered data location as secondary. In this section, we have a look at some of the important scheduling research, which has considered the importance of data location while making the scheduling decision.

Reference [61] describes an adaptive scheduling algorithm for data grids. It describes a grid broker that mediates access to distributed resources including data. The responsibility of a resource broker is to hide the complexity of the grid by transforming user requirements to jobs that can be scheduled on the most appropriate resources. The result of the execution is then returned to the user. The resource broker has been termed a gridbus broker. It provides services suitable for data-intensive applications. The scheduler within the gridbus broker makes scheduling decisions considering the *network proximity* of data from the computing resources. Network proximity can be defined in terms of the available bandwidth between the data source and computing resources. A network monitoring service is used to obtain this information. Based on the parameters supplied by the user, the broker obtains a list of *logical data files (LDF)* from a grid catalog service. These files are then mapped to actual hosts and a list of data and computational resources for the user's job is obtained. A data-compute resource pair is selected for the job that can complete the job earliest. This decision is made by considering the estimated execution time of the job on the machine, data transfer time using current bandwidth estimates and the job completion rate of the machine.

An algorithm for scheduling *parameter sweep application (PSA)* in grid environments is described in [62]. A parameter sweep application is an application that executes multiple instances of a program using different sets of parameters and then collects the results from all the instances. Such applications frequently arise in scientific and engineering problems. PSAs are independent, i.e., they do no communicate among themselves. However, distinct experiments in PSA share large input files and produce large output files. So, for the efficiency of the parameter sweep experiments, the experiments and data must be located near to each other. The basic approach is to place the input files strategically so that they can be used by a maximum number of PSA experiments. A way of doing this could be to stage the file in a storage cluster and schedule the experiments to the nodes near it. PSA employs three heuristics for mapping file transfer to network links and computation to nodes. The three heuristics used are: min-min, max-min and sufferage. These heuristics are slightly modified to include the data transfer time for input and output files while computing the minimum completion time (MCT). A sufferage algorithm works by finding the best and the second-best MCT for a task. The sufferage value is defined as the difference between the best and second-best MCT. The tasks are scheduled in decreasing order of the sufferage value. The heuristics behind using the sufferage value is that the task that would *suffer* the most is assigned to execute first. The same idea is used for the placement of jobs in clusters that already contain the necessary input files. If the task is not placed on a node in the cluster, the task will suffer because the input file will have to be copied from the cluster. However, the sufferage heuristic does not perform well in this scenario. This is explained as follows. Suppose a cluster A contains a large input file required by a task. If the cluster contains two or more nodes having almost identical performance for that task, they would have very similar MCTs for the task, resulting in a sufferage value tending to zero. The task will then have low priority and other tasks will be scheduled ahead of it. This may force the task to be scheduled on another cluster B requiring the transfer of input file from cluster A to cluster B. This may severely affect the performance of the task. So, a modification is made to the standard sufferage heuristic and is called the XSufferage heuristic. In XSufferage, for each task, the task's MCT is calculated for each cluster called the cluster-level MCT. The cluster-level sufferage value is computed by finding the difference between best and second best cluster-level MCTs. The task that has the highest cluster-level sufferage value is scheduled on a host that provides the minimum MCT within the cluster that provides the earliest cluster-level MCT. Consider a task T, which is being considered for execution on two clusters: cluster A and cluster B. Cluster A has the necessary input files for T. The cluster-level MCT for the task T executing at cluster A will be low because the *intra-cluster* data transfer time is much less. On the other hand, the cluster-level MCT for the task T executing at cluster B will be driven by the *inter-cluster* data transfer time for copying the input files from cluster A to cluster B. This will result in a high cluster-level sufferage value for task T and so it will be given priority and scheduled at a node in cluster A. So by modifying the sufferage heuristic and incorporating the data transfer time in MCT, a task can be scheduled in the cluster having the necessary input files.

Reference [63] explains a scheduling strategy for data-intensive applications in which the data movement may be *tightly* coupled or *loosely* coupled with the scheduling decision. By tightly coupled we mean that the scheduling decisions take into account the necessary data movement operations. Loosely coupled implies that the data movement can be performed in an asynchronous way based on load and data access pattern. The system model consists of three schedulers.

- *External Scheduler (ES):* Each site contains an ES where the users submit their job. It determines the remote site on which the user job should be submitted. For this the scheduling algorithm used by ES needs to know the load at the remote site and the location of the data.

- *Local Scheduler (LS):* Once a job is submitted to a remote site by the ES, it is the responsibility of the LS to schedule these jobs on the local resources.

- *Dataset Scheduler (DS):* The DS keeps track of the *popularity* of the locally available data (datasets that are frequently accessed are replicated at remote sites). Before replication, the DS needs to know whether the data is available at the remote site.

ES selects a remote site using one of the four algorithms: JobRandom, JobLeastLoaded, JobDataPresent and JobLocal. *JobRandom* randomly selects a site for executing the job. *JobLeastLoaded* selects a site having minimum load. Reference [63] defines load as the number of jobs waiting to run. *JobDataPresent* schedules the job to a site that already has the data. In case of tie, one site is chosen randomly. JobLocal runs the job locally irrespective of the data location and local load.

DS defines three algorithms: DataDoNothing, DataRandom and DataLeastLoaded. *DataDoNothing* does not adopt any dynamic replication policy. If data needs to be accessed from a remote site for a job, it is maintained as a cached dataset (or files) using *Least Recently Used (LRU)* policy. *DataRandom* keeps track of the popularity of the dataset it contains. When the popularity exceeds a certain threshold, the dataset is replicated to a random site. *DataLeastLoaded* selects the least loaded neighbor site for replicating a popular dataset. Based on the algorithms used by ES and DS there are 12 possible combinations of the algorithms. The strategy JobDataPresent used with DataRandom or DataLeastLoaded gives the best performance. This combination can be described as *scheduling of jobs at the data sources or at the least loaded actively replicated popular data*. The reason for its best performance among other combinations of algorithms is that it ensures load sharing in the grid and requires minimum data transfer. To sum up, good performance is obtained by scheduling the job at data sources and by generating new replicas of popular datasets at each site periodically. The importance of the approach discussed in [63] is that it decouples the job scheduling and data replication strategy contrary to approaches that try to generate a globally optimal schedule for both. Decoupling allows optimization of data replication and job scheduling separately, without considering the impact of the other. This allows a decentralized and thus scalable implementation of scheduling strategy, which is desired in grids.

3.3   Service Monitoring and Discovery

When a job is submitted to the grid for execution, the scheduler maps the job to the suitable resources using mapping heuristics. For this mapping to be possible, the grid needs to be aware of the existing resources and their status (available or busy). The resources in a grid are very varied, ranging from computers, networks, storage and devices to protocols and algorithms used by an application. Monitoring is the process of observing the resources and services to find their usage patterns or to find faults. Discovery is the process of finding suitable resources to perform a task. In a dynamic environment like grids, where resources can join or leave the system without informing, an efficient service discovery mechanism is essential for mapping the tasks to the best possible resources in the grid environment. The service discovery is complicated by the heterogeneity, dynamic nature and geographical distribution of the resources in the grid. The requirements of a grid information service is largely motivated by the requirements of the grid. The information sources in grids are geographically distributed and are subject to failure. The number of information sources can be very large ruling out the possibility of a centralized information service.

We will see in the subsequent sections that grid information services use a hierarchical architecture to provide fault tolerance and cope with the large number of users. Information about a resource in the grid can become stale due to change in the status of the resource. This necessitates maintaining a time-to-live metadata along with the information about the resources. Another requirement associated with the information system is to *not* provide consistent information about the resources to the users. This requirement stems from the fact that a grid information system is distributed and providing a consistent global state of the grid may be very expensive and does not scale well. If an application requires consistent global state, this functionality can be achieved through other control functions at a higher cost [35]. In a distributed environment, both resources and information providers about those resources are bound to fail. A grid information system should be resistant to such failures. Suppose a part of the virtual organization (VO) goes down due to network failure. This should not prevent grid users from obtaining information about resources in other parts of the grid. Further, the information system should have timely information about failures. This is achieved through a *soft-state model*, which means that information about a resource may be discarded unless it is refreshed using a timely notification from the information provider. So in our example, if the information provider becomes disconnected from the grid due to network failure, it won't be able to provide information about the resources to the grid information service. When the time-to-live information associated with the resources expire, the grid information service purges those resources assuming that the resources are temporarily unavailable. When the network starts functioning again, the information provider provides information about the resources to the grid information service and those resources again become available to the grid users. A grid information service may provide information about resources located in different VOs. So there might be restrictions on the dissemination of information. Information about resources should be provided only after validating the identity of the requester and whether the requester has sufficient privileges to the information requested. This requires a strong authentication and authorization mechanism associated with the grid information service. Other desirable features of a grid information system are: caching to improve the performance of information retrieval, standard representation of resources, extensibility, expressiveness to represent relevant information in the grid environment and easy deployability [36].

3.3.1   Grid Information System

Having discussed the requirements of a grid information system, we now discuss two grid information systems: (a) Globus Toolkit Monitoring and Discovery Service (MDS2) and (b) European Data Grid Relational Grid Monitoring Architecture (R-GMA). As we discuss these information systems, you will be aware of the finer details of the requirements of the information system in the grid and how they are implemented.

3.3.1.1   Monitoring and Discovery Service (MDS)

The Monitoring and Discovery Service (MDS) of the Globus Toolkit provides resource monitoring and discovery services within the grid environment.

In this section we talk about the protocols used within MDS and the components that define the hierarchical nature of MDS. When we discuss the Grid Data Model in section 3.3.3, we will see how MDS2 makes use of *LightWeight Directory Access Protocol (LDAP)* to define an extensible data model that can be used to represent the information about entities in the grid and how it provides a scalable implementation, well suited to the distributed environment.

MDS has a hierarchical structure. It consists of three main components: Information Provider, GRIS and GIIS. An *information provider (IP)* is defined as a service that interfaces a data collection service and provides information about the available resources to *Grid Resource Information Service (GRIS)*. The GRIS collects information about the resources from the IPs and makes the information available to the *Grid Index Information Service (GIIS)*. The GIIS is an aggregate directory of the information available at the lower level. The GRIS registers itself to the GIIS, which in turn requests information from GRIS. The services register with other services using a soft-state protocol thereby allowing automatic and dynamic cleaning of unavailable or dead resources. Caching is used at each level to keep information about static resources. This decreases the network overhead.

An IP speaks two protocol [35]: (a) *Grid Information Protocol (GRIP)* and (b) *Grid Registration Protocol (GRRP)*. GRIP is used to access information about entities in the grid. It supports both *discovery* and *enquiry* about the resources in the grid. Discovery is supported through a search capability whereas enquiry refers to direct lookup of information. The resource name is supplied in enquiry and the resource description is returned. MDS uses LDAP as the protocol for GRIP. LDAP defines a *query language* for facilitating search and lookup. A grid entity can contain many attributes. A filter can be specified by the query language to obtain information about only a subset of those attributes. LDAP also supports *query-reply*, which consists of attributes from matching objects (as specified by the filter). LDAP query language does not support relational joins [35], which allows several different entities to be merged as a single entity. GRRP allows one component (service) of MDS to inform other components about its existence. IPs use GRRP to notify GRIS about their availability. It is also used by an aggregate directory like GRIS to invite an IP to join a particular VO. Local and VO-specific policies determine which aggregate directory an IP should register itself with. As mentioned, the components of MDS use soft-state protocol for communicating among themselves, the aggregate directory can remove an IP from its list if it has not received a registration message from the IP for a certain duration.

MDS provides security features through the use of *Grid Security Infrastructure (GSI)*. GSI (4) uses X.509 digital certificates for the purpose of authentication. So, mutual authentication between information provider and consumer can be carried out by using X.509 digital certificates. After the authentication process is complete, the information provider may be queried about the resources. The information provider may limit the extent of information available to the aggregate directory, based on the local policy. Information such as operating system type of available physical memory may be available to the aggregate directory. But information such as the current system load or available network bandwidth may be made available to only selected users. The message exchange used during registration protocol (GRRP) also needs to be secured. This can be done either by using transport layer security or by encrypting the message using the X.509 digital certificate of the registering entity. The aggregate directory can make a decision regarding the selection or rejection of registration requests made by IPs.

3.3.1.2 Relational Grid Monitoring Architecture (R-GMA)

Relational Grid Monitoring Architecture (R-GMA) [37] has been developed by the European DataGrid project as an information and monitoring system for grids. It is based on the Grid Monitoring Architecture (GMA) [38], which is based on a simple consumer-producer model and meets the requirements of a distributed information system. R-GMA derives its strength and flexibility from the relational model. GMA consists of three components: a producer, a consumer and a directory service. The producers register themselves with the directory service and describe the type and format of information that they would provide to the grid. A consumer queries the directory service to know the type of information available and the producers that provide such information. The consumer can then directly communicate with the producer to obtain the necessary details. Current GMA definition does not mention the underlying protocols and data models to be used. GMA also allows registering consumers to the directory service, so that the producer can find the consumer. The GMA definition does not mention any protocol for the underlying data model. R-GMA, which is based on GMA, uses RDBMS to store information about the resources in the grid. Producers use SQL `createtable` statement to provide information about the format of data being published. The data is published using SQL `insert` statement. Consumers make use of SQL `select` statement to specify the criterions for a search.

R-GMA enables creation of a *virtual database*, which can be queried by users to obtain information about resources in the grid. As the data is organized into tables, users can query and insert data using SQL constructs. The virtual database consists of the schema (list of the table definitions), registry (consisting of a list of data providers) and a set of rules for deciding the data providers to be contacted for a particular query. *Mediator*, a component encapsulated in the consumer interface [37], contains these rules. There is no constraint on the number of virtual databases in a grid, provided each has a unique name. Generally, a virtual database is owned by a VO.

R-GMA defines five types of producer interfaces [37]:

- DataBaseProducer: It writes each record to an RDBMS and so it can handle joins.

- StreamProducer: It writes information to a memory structure that can be read by the consumer. The consumer can subscribe to a stream having specific properties directly from the StreamProducer.

- ResillientProducer: It is similar to the StreamProducer except for the fact that the data is backed up at a secondary storage to prevent information loss during system crash.

- LatestProducer: It holds only the latest records. Each record is associated with a timestamp, a set of fields that define what item is being measured and a set of fields defining the actual measurements. Latest-Producer replaces an earlier record having the same set of defining fields and a timestamp earlier than the newer one.

- CanonicalProducer: It is different from other producer types. It does not provide an interface to publish data using an SQL insert statement, like the other producers. It triggers a user code in response to an SQL query.

R-GMA is based on the servlet technology. At the time of writing this book, development of an OGSA R-GMA framework was under process.

3.3.2 Aggregate Directory

The aggregate directories provide a collective repository for resources present in the grid. They can provide a specialized view of resources and services in a grid [35]. For example an aggregate directory intended for supporting application monitoring might keep a list of the running applications. A directory used by a mapper for making scheduling decisions may keep a list of machines available to a VO, organized by their machine architecture types. Such specialized directory services may keep indices (just like databases), to improve the response time for such specific queries.

In the previous section we talked about two grid information systems, MDS and R-GMA. MDS uses a name-serving aggregate directory, which contains a record for all the entities registered to the directory. It supports queries based on name resolution. R-GMA uses a relational database to store information about entities in the grid that can be queried using SQL statements.

The organization of an aggregate directory can have a significant impact on the performance of the Grid Information Service. If there is only one aggregate directory for every VO that maintains the list for every entity in the

VO, the design becomes non-scalable. As the number of resources in the VO increases, the amount of data stored by the aggregate directory increases. Further, as all the queries are directed to the same directory service, it decreases the response time for the grid users. A single aggregate directory also means a single point of failure. To overcome these issues, a hierarchical structure has been adopted for the aggregate directories. The hierarchical structure of the aggregate directories also reflects how a VO administration is typically handled. A VO may consist of several organizations. Each of these organizations maintains a local aggregate directory. These organizations can then register their local aggregate directory to the directory maintained by the VO. An organization may be part of several VOs. So it may need to register the local directory with directories maintained by all the VOs. So we see that the hierarchical organization of the aggregate directories serves two purposes. It provides scalability and fault-tolerance. It facilitates easier administration of directory services based on the administrative policies of each site participating in the VO.

Monitoring and Discovery Service of the Globus Toolkit is based on the hierarchical organization of aggregate directories, as we just described. MDS uses two kinds of nodes for aggregate directories: Grid Resource Information Service (GRIS) and Grid Index Information Service (GIIS). The GRIS registers itself to one or more GIISs. The GIIS then requests information about the resources from GRIS. The GIIS can in turn register to one or more GIISs higher in the hierarchy. The hierarchical structure of MDS directories is illustrated in the figure 3.1.

MDS2 uses a *pull* model for data exchange. This means that a GIIS, for example, need not explicitly store the information about the entities. When a user requests information about a grid resource, it can be dynamically generated by querying a lower level aggregate directory (for example a GRIS). On the contrary, iGrid [39], a grid information service developed within the European GridLab project, uses the *push* model for data exchange. iGrid defines two kinds of nodes for its hierarchical directory structure: *iServe* (similar to GRIS) and *iStore* (similar to GIIS). iServe periodically sends information about the resources to the registered iStores. So, unlike MDS2, where a GIIS may need to query lower level directories for information about a resource, an iStore is guaranteed to have updated information.

3.3.3   Grid Information Service Data Model

MDS extends the LDAP data model to represent information about entities in the grid in a flexible manner. The MDS data model consists of specialized object definitions to represent diverse entities in the grid environment. It allows specifying the properties of computers as well as the network and the connections between computing resources. The data representation and the

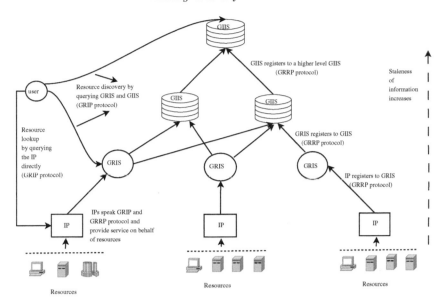

FIGURE 3.1: Hierarchical structure of MDS.

application programming interface for the MDS has been adopted from LDAP [36]. In this section we explain the naming convention used by MDS and describe some of the important object types in the MDS data model.

An MDS entry is identified by a unique name, called its *distinguished name*. The MDS entries are organized as a hierarchical, tree-based structure called the *Directory Information Tree (DIT)* [36]. The distinguished name for an entry is generated by enumerating the nodes on the path from the root to the entry being named. Each path, from the root to any node in the tree, should generate a unique name. This requires that for every DIT node, its children must have at least one attribute that distinguishes it from its siblings. The components of the distinguished name are listed in *little-endian* order, with the root of the DIT listed last. An example of MDS distinguished name can be:

```
< hn = node1.lisa.ecp.fr,
  ou = LISA,
  o  = Ecole Centrale Paris,
  o  = Globus,
  c  = France >
```

We see that the country entry, which is at the root of the DIT, is mentioned at last. A pseudo-organization named Globus has been defined by MDS, so that name defined for information services does no clash with names defined for other purposes [36].

Every DIT entry belongs to an object class that determines the attributes that entry may have and also the values those attributes can take. The definition of an object class consists of three parts: a parent, a list specifying the required attributes and a list specifying the optional attributes [36]. The object class may contain a SUBCLASS section, which allows an object class to inherit attributes from other object classes. So the SUBCLASS section, provides support for the inheritance mechanism. The MUSTCONTAIN and MAYCONTAIN sections list the required and optional attributes in the DIT entry. Each attribute name is followed by a type of value it can take. The value of an attribute within a DIT entry may be another DIT entry. This enables construction of complex relationships that cannot be depicted by the DIT.

Having described the elements of object classes, we now describe the MDS data model. The computing nodes and people are the immediate children of the organization in the DIT tree. Nodes represent the object class GlobusHost, which is a subclass of the GlobusResource class. The following [36] shows the simplified versions of GlobusHost and GlobusResource object classes (*cis* stands for case insensitive string and *dn* stands for distinguished name).

```
Globus Host OBJECT CLASS
     SUBCLASS OF GlobusResource
     MUST CONTAIN {
          hostname              :: cis ,
          type                  :: cis ,
          vendor                :: cis ,
          model                 :: cis ,
          OStype                :: cis ,
          OSversion             :: cis
     }
     MAY  CONTAIN {
          networkNode           :: dn ,
          totalMemory           :: cis ,
          totalSwap             :: cis ,
          dataCache             :: cis ,
          instructionCache   :: cis
     }


     GlobusResource OBJECT CLASS
     SUBCLASS OF top
     MUST CONTAIN {
          administrator         :: dn
     }
     MAY CONTAIN {
          manager               :: dn ,
          provider              :: dn ,
          technician            :: dn ,
          description           :: cis ,
          documentation         :: cis ,
     }
```

The networks are also immediate children of the organization. To represent an entry of type communication network, MDS uses the GlobusNetwork object class. Its attributes provide information about the link protocol (ATM or ethernet), network topology and the physical media. Networks and hosts are related to each other via the `GlobusNetworkInterface` objects. They are located under GlobusHost objects in the DIT. If a node has more than one network interface, it has one network interface entry for each of them in the GlobusHost object. GlobusNetworkInterface object represents the physical characteristics of the network interface such as the interface speed and the hardware network address. GlobusNetworkInterface objects are associated with the corresponding GlobusNetwork object through an attribute in the GlobusNetworkInterface, whose value is the distinguished name for the corresponding network object. The network object contains a reverse link that points to the network interface.

To represent logical information, such as network protocols associated with hosts and networks, MDS uses the concept of *image* classes. `GlobusHostImage`, `GlobusNetworkImage` and `GlobusNetworkInterfaceImage` objects are the image classes for corresponding physical objects. These objects contain information associated with the logical view as well as a distinguished name pointing to the corresponding physical object.

3.3.4   Grid Service Discovery

To find information about a resource, the user can either directly query the resource or query an aggregate directory. For example a client can search for a resource from the GRIS directly. To obtain a more comprehensive and collective list of resources, the user needs to query the GIIS. Another way of doing this is to query an aggregate directory like GIIS for the location of a service and then query the corresponding service directly to obtain detailed information about it. The information becomes stale as we move up in the MDS aggregate directory hierarchy. This happens because information with an aggregate directory at a higher level is propagated upwards by the Information Providers (IPs) and lower level aggregate directories. This information is refreshed after a certain duration. During this duration the actual state of the resource may have changed (for example, from busy to free). This information is not available immediately at the higher aggregate directories. So the closer the aggregate directory is to the information source, the fresher is the information available with it. At the same time, an aggregate directory at a lower level in the hierarchy may not have comprehensive list of available services.

Now we talk about the services provided by MDS4 for monitoring and discovery. MDS consists of an aggregator framework, which is a software framework used to build services that collect and aggregate the collected data. An aggregator framework collects data from *aggregator sources*, which are Java classes that implement an interface to collect XML-formatted data. The services built on top of the aggregator framework are called *aggregator services*. These services use a common configuration mechanism to maintain information about the aggregator sources from which data has to be collected. The configuration file also maintains the list of parameters supported. As mentioned earlier, each registration has a lifetime. If the registration is not renewed, the associated entry and the data are removed from the server. MDS4 supports three aggregator services [40] that collect information from registered information sources. The aggregator services have different interfaces and behaviors. These are:

- MDS-Index: It allows XPath queries on values obtained from the information sources.

- MDS-Trigger: If the collected information matches the user-supplied criteria, a user-specified action is triggered (for example, entry to a log file).

- MDS-Archiver: It stores information obtained from the information sources to a persistent storage from where they can be queried by the user to obtain historical information.

So to sum up, aggregator framework collects information from aggregator sources (which in turn collect information from information providers) and this information can then be used by aggregator services to answer client queries. MDS4 uses XML and Web Service for registering information sources and finding resources of interest. The information collected by the three aggregator services are stored as XML files. These can be queried via XPath queries [40]. XPath queries are more efficient than LDAP queries used in earlier versions of MDS (MDS2). MDS4 supports registration of information sources via WSRF-compliant web services and supports both *push* and *pull* subscription methods.

3.4 Grid Workflow

If we ignore grid computing for a moment and look at the definition of workflow, it may be defined as an automation of a process, consisting of many participants (called sub-processes), where information or data may be passed between these participants based on a predefined set of rules. Now we look at workflow from a grid perspective. A task in a grid may be composed of several sub-tasks, which often have dependency among themselves. For example, one sub-task A may need the result from sub-task B. So sub-task A should be scheduled to execute after the completion of sub-task B. Another example may be two sub-tasks using parallel execution may during their execution communicate with each other using remote procedure calls. If we consider web services, which have become an integral part of grids with the advent of Globus Toolkit 4.0, a new web service may be formed from two or more web services. In this case, again, workflow defines the order of execution of the web services within the new web service and the messages exchanged among them. From all three examples, we can think of grid workflow as a sequence of several different activities or services that have to be combined to solve a specific problem or to create a new service. This process of *automation* of a collection of services to generate a new service in the grid, may be termed as grid workflow. Grid workflow also requires coordination of the services through control and data dependencies. So a more complete definition of grid workflow would be the automation of a collection of services by coordination through control and data dependencies to generate a new service.

Grid computing is being widely used in scientific communities such as bioinformatics, astronomy, high energy physics, etc. Such applications require orchestration of diverse grid resources and executing complex workflows. Grid workflow offers several advantages [41]. It allows the building of dynamic applications from heterogeneous distributed resources. Grid workflow enables management of applications that require resources from different administrative domains.

Grid workflow can be categorized as linear, acyclic or cyclic [41]. In a *linear* workflow tasks have to be performed in a specified linear sequence. For example, we provide a data set as input to task A, which performs some processing on it and passes the modified data set to task B, which manipulates the data set in a different way and passes it to task C and so on. Such grid workflows are very simple to define and can be described using a simple scripting language. More complex grid workflows can be represented as *acyclic graphs* where nodes represent the tasks and edges represent the dependencies between the tasks. It is difficult to represent such workflows using scripting languages and they require the use of workflow languages. *Cyclic graphs* take the grid complexity to the next level. Communication among services in a grid often take the form of a series of message exchanges. Edges in a cyclic graph represent services that are connected to each other through a series of message exchanges. In the subsequent sections we talk about the workflow languages, grid workflow management systems and the workflow scheduling algorithms.

3.4.1   Grid Workflow Management System (GWFMS)

A *Grid Workflow Management System (GWFMS)* deals with modeling tasks and their dependencies in a workflow and managing execution of workflow and interaction with grid resources. As mentioned in [41] it consists of five components: workflow design, information retrieval, workflow scheduling, fault tolerance and data movement. Functions of a GWFMS are broadly classified into build-time functions and run-time functions. *Build-time* functions involve modeling tasks and dependencies among them using workflow design and definitions. *Runtime* functions include the execution and control of workflow. After the user has generated the workflow using modeling tools, it is submitted to a run-time service called the grid workflow enactment service. This service interacts with grid middleware to provide services like workflow scheduling, data movement and fault tolerance. During build-time and runtime stages the GWFMS needs to interact with the grid information services to find the status of grid resources. Now we discuss the components of GWFMS in detail.

3.4.1.1 Workflow Design

A workflow design consists of four key components [41]: workflow structure, workflow specification, workflow composition system and workflow QoS constraints. As explained in the grid workflow definition, a workflow consists of multiple tasks having dependencies among them. A workflow structure indicates the temporal relationship among these tasks. Workflow structures may be represented using either a DAG (Directed Acyclic Graph) or a non-DAG. DAG workflows may be classified as *sequence*, *parallel* and *choice* based on the ordering of tasks in the workflow. In sequence workflow structures there is a strict ordering among the tasks and one task starts execution only after the previous task has completed. Parallel workflow is characterized by two or more tasks executing concurrently and possibly communicating among themselves. In choice workflow structures, the decision to execute a task may be made dynamically, if a condition is met. Non-DAG workflow contains all the patterns listed in a DAG-based workflow. In addition to those, it also contains an *iteration* pattern. In iteration workflow, a subset of a workflow inside an iteration loop can execute more than once based on the loop counter or a predefined condition. Workflow structure can also be modeled using High-Level Petri Nets (HLPN). We will discuss workflow structure using Petri-nets when we talk about workflow specification languages.

Workflow Specification provides the task definition and structure definition for a workflow. Workflow specification, also termed workflow model, can be categorized as abstract or concrete. In an *abstract* specification, the grid resources required for task execution are not stated. Tasks can be mapped to suitable resources at the run-time using the grid information discovery services. A *concrete* workflow specification associates tasks to appropriate grid resources before hand. As the availability of resources in grids is very dynamic, it is generally more appropriate to define the workflow specification as abstract. During the actual execution of the workflow a concrete specification may be generated based on the status of the resources [41].

The workflow composition system allows users to develop a grid workflow from individual components and hides the underlying grid complexity. Reference [41] presents a comprehensive taxonomy for the workflow composition systems. They can be broadly classified into user-directed composition systems or automatic composition systems. *User-directed* systems allow the user to specify the workflow either using workflow specification languages (called *language-based modeling*) or using tools to generate the workflows automatically for the user (called *graph-based modeling*). Graph-based modeling provides a GUI where a user can drag-and-drop a required component into the workflow diagram. Language-based modeling techniques include languages such as XML, WSFL, GridAnt, XLANG, etc. Graph-based modeling includes tools such as UML and petri nets.

A grid may consist of multiple resources that can fulfill the requirements of a task. However these resources may vary in terms of their QoS specifications. So a workflow design must include QoS contraints because different users have different requirements. A user should be able to specify his QoS expectation in the design of the workflow. Workflow QoS constraints may include time, cost, reliability, etc.

3.4.1.2 Information Retrieval

A grid workflow management system coordinates execution of tasks by mapping the tasks to suitable resources. To do this, one must query suitable information sources. These information sources can be classified by the type of information they provide [41]: static, historical and dynamic. *Static* information refers to information that does not change with time. This includes information about the infrastructure, such as number of processors, or configuration-based information such as link-layer protocol, OS version, etc. Historical information is obtained from records of previous application execution by the grid. *Dynamic* information refers to information about those grid components that keep changing with time, for example, CPU usage, available network bandwidth, free disk space, etc. GWFMS makes use of Grid Information Services to obtain information about the resources in the grid. MDS either directly provides information about the grid resources or can make use of another service to obtain the required information.

3.4.1.3 Workflow Scheduling

Workflow scheduling in grids can be categorized as local scheduling and global scheduling. *Local scheduling* refers to scheduling of tasks on a local cluster or providing time slots to tasks on a single machine. *Global Scheduling* means determining which resource to allocate for the execution of a task on the grid scale. Workflow scheduling refers to global scheduling. The resources in a grid are under various administrative control and policies. So a workflow scheduler needs to coordinate among the diverse local management policies. The workflow scheduler also takes into account the user's QoS requirements while making the scheduling decision. A workflow scheduling architecture may be centralized, hierarchical or decentralized [41]. In a *centralized* architecture, a single scheduler makes scheduling decisions for all the tasks in the workflow. The scheduler has the complete information about the workflow and the resources pertaining to it. However, such a design does not scale well in practical grid scenarios. It is suitable in scenarios where the scheduler is dealing with a small workflow with respect to the number of tasks or all the tasks in the workflow require access to the same set of resources. In a *hierarchical* scheduling architecture, there is a centralized scheduler with control of multiple lower-level schedulers, which schedule tasks in the sub-workflow. The top-level scheduler assigns sub-workflows to the lower-level schedulers. The lower-level scheduler schedules tasks in the sub-workflow to resources owned by one organization. This architecture allows different scheduling policies to be deployed at the local level and at the central level. The hierarchical scheduling architecture is susceptible to a single point of failure. In a *decentralized* scheduling architecture, there is no central point of control and different schedulers that schedule tasks in a workflow may communicate among each other. However, conflicts may arise when tasks in different sub-workflows compete for the same resource. Although this design is scalable, it may lead to a sub-optimal scheduling of the workflow.

The decision to map tasks from a sub-workflow can either be local or global. In local decision only those tasks from a single sub-workflow are considered. Although this method results in optimal sub-workflow schedule, it may reduce the overall performance of the workflow. For example a local scheduling decision may not consider the data-transfer between tasks in a different sub-workflow. For global scheduling decisions graph-based algorithms can be used that analyze the complete graph and find conflicts among parallel tasks competing for the same set of resources. Global scheduling decisions take more time, and for applications that are not data-intensive a local scheduling decision should suffice.

Workflow scheduling involves a planning scheme that involves mapping abstract workflows to concrete workflows. The schemes can be static or dynamic [41]. In a *static scheme*, the mapping is done based on the knowledge of the current execution environment and dynamic nature of the resources is not considered. A *dynamic scheme* takes into account both the static and dynamic nature of the execution environment. Static schemes may involve intervention by the user to map the tasks to resources before the actual execution. It can also be done by simulating the execution of workflow on a set of resources before the actual workflow execution begins. Dynamic schemes involve predicting the execution of the workflow on a set of resources. An initial scheduling scheme may change if a better resource is found for execution. Workflow scheduling algorithms will be covered in Section 3.4.3.

3.4.1.4 Fault Tolerance

A workflow may fail due to non-availability of a resource or software component, network failure, system overloading and modification in the system configuration violating the initial contract. A GWFMS should identify such failures and take necessary corrective measures. Reference [42] classifies workflow fault handling techniques into task-level and workflow-level fault handling. *Task-level* techniques try to take corrective measures at the task level. For example they may try to restart the task on the same set of resources or may allocate the task to a different set of resources. Task-level techniques also involve checkpointing the task so that they can resume execution from the point of failure, possibly on another resource. Replication techniques [42] run the same task on a different set of grid resources. So if the task can be completed successfully on even one of the resources, the execution is successful. *Workflow-level* fault handling manipulates the structure of workflow to deal with error conditions. An example of such a technique is the DAGMan [43] meta-scheduler for Condor [44], which ignores the failed task and continues the execution of the remaining workflow until no forward progress can be made. The DAGMan then generates a rescueDAG, a rescue workflow description, which indicates the node failures along with statistical information.

3.4.1.5 Intermediate Data Movement

If an application requires input data from a remote site, the GWFMS needs to stage the data files before application execution. Similarly, after the execution has completed, the result of execution needs to be transferred to the site that requested the execution. Further, there might be a task that requires output data from another task in the workflow and both tasks have been scheduled to run on different resources. For such cases also, the GWFMS needs to transfer the necessary data files. For small workflows, data movement can be carried out by the user. However, for large workflows, automatic data movement is preferred. A *centralized* approach may be used for data transfer, in which a central manager can collect intermediate data from a a task and forward the data to the task's successor in the workflow. In a *mediated* approach, the intermediate data generated by the tasks are managed by a distributed data management system. The party needing the data can retrieve it from the data management system. In a *peer-to-peer* approach, the data transfer does not involve a third party and can be directly transferred between source and destination nodes. The peer-to-peer approach requires every node in the grid to have data management and data movement capability.

3.4.2   Workflow Specification Languages

Workflow specification languages define a formal way to express the causal
or temporal dependencies among a number of tasks to be executed in the
grid. Workflow languages use XML notation to represent inter-task dependencies. Most of the workflow languages fall under one of the three categories:
script-based workflow languages, purely graph-based workflow languages and
Petri-nets-based workflow languages. Examples of script-based workflow descriptions are GridAnt and BPEL4WS. They contain constructs for workflow
such as sequence, parallel and while/do. These are difficult to use for unskilled
users. Purely graph-based workflow descriptions include Condor's DAGMan
tool and Symphony. These workflow languages are based on Directed Acyclic
Graphs (DAGs). In DAGs, communication between services are represented
as arcs going from one service to another. However, it is difficult to express
complex workflows, like loops, using DAGs. Workflow languages based on
high-level Petri-nets (HLPN), such as GWorkflowDL, provide greater flexibility than DAGs. HLPNs provide extensions to classical Petri-nets by adding
support to model data, time and hierarchy. Petri-nets model the state of
program execution by using the concept of tokens flowing through the net.
Petri-nets can easily model the data flowing through the nets, which is difficult to model using DAGs. The advantage of using Petri-nets for workflow
management is that they can be verified for the correctness of workflow and
checking conditions such as deadlock and liveness. We now describe three
workflow specification languages: GridAnt, GWorkflowDL and GSFL.

3.4.2.1   GridAnt

GridAnt [45, 46] is an extension of the Apache Ant build tool. It allows
construction of client-side workflows for Globus applications. It contains constructs for expressing parallel and sequential tasks. The core of the GridAnt
system lies in the workflow engine, which is responsible for orchestrating control and data through multiple GridAnt activities. As it is based on Ant,
which is written in Java, it provides platform-independence and seamless integration in the grid services framework. Further, Ant is highly extensible,
making it easier to incorporate grid-specific functionalities [46]. However,
Ant does not provide support for conditional execution, error and exception
handling of arbitrary control flows that are required by any practical grid application. It also lacks the ability to allow the output of one activity to become
input of another. To overcome these limitations of Ant for grid environments,
GridAnt was developed to support workflow orchestration and composition in
grid environments. GridAnt also provides a visualization tool that loads the
XML workflow specification into a hierarchical graph. For example, it transforms the dependencies between GridAnt targets into nodes connected by an
edge. GridAnt provides a workflow vocabulary, which is a set of pre-defined
tasks. They can be used to construct more complex workflows. Following is
the list of important (but not comprehensive) tasks outlined by the GridAnt

framework [46].

- *grid-setup*: It establishes the grid environment based on client preferences.

- *grid-authenticate*: It initializes the system for subsequent authentication. For Globus, a proxy credential is created to be used for single sign-on for a limited lifetime.

- *grid-execute*: It executes the task on a remote grid resource.

- *grid-copy*: It copies a file from one node in the grid to another. Third-party transfers are also supported through the use of GridFtp protocol.

- *grid-cancel*: It cancels the execution of the given job on a remote machine.

- *grid-query*: It queries the grid information service to find the status and capability of a grid resource.

- *grid-status*: It queries the remote resource to know the execution status of a given job.

We now provide examples of sequential and parallel constructs used in the GridAnt specification language. Tasks encapsulated in `<sequential>` `</sequential>` are executed in sequential order.

```
<sequential >
    <grid-execute name = "Find Differential">
        environment = "PATH=/home/user1"
        directory = "/home/user1/diffProject"
        protocol = "GRAM"
        server = "node1.lisa.ecp.fr:2135"
        executable = "findDifferential"
        output = "out.txt"
    />
    <grid-copy from = "gridftp://node1.ecp.fr:/home/user1
        .../diffProject/out.txt"
            to = "gridftp://node2.lisa.ecp.fr:/home/user1/
                ...Results/result.txt">
</sequential >
```

This example results in sequential execution of two tasks. First it performs the differentiation of an expression and stores the result in out.txt. Then that file is copied to another node using GridFtp protocol. Similarly, tasks encapsulated in `<parallel></parallel>` are executed in parallel fashion:

```
<parallel >
    <grid-copy from = "gridftp://node1.lisa.ecp.fr:/home/
        ...user/a.txt"
            to = "gridftp://node2.lisa.ecp.fr:/home/user/b.
                ...txt">
    <grid-copy from = "gridftp://node3.mas.ecp.fr:/home/
        ...grid1/d.txt"
```

```
                    to = "gridftp ://node2.mas.ecp.fr:/home/grid/e.
                        ...txt">
        </parallel>
```

The above example copies two files using GridFtp protocol concurrently. Dependencies in GridAnt are described by using the **target** construct.

```
        <target name = "getFirstMatrix"
            description = "Get file1 containing the first matrix
                ...">
            <grid-copy from = "gridftp ://node2.lisa.ecp.fr:/home/
                ...user/file1.txt"
                to = "gridftp ://node1.lisa.ecp.fr:/home/user/
                    ...file1.txt">
        </target>
        <target name = "getSecondMatrix"
            description = "Get file2 containing the second matrix
                ...">
            <grid-copy from = "gridftp ://node3.lisa.ecp.fr:/home/
                ...user/file2.txt"
                to = "gridftp ://node1.lisa.ecp.fr:/home/user/
                    ...file2.txt">
        </target>
        <target name = "matrixMultiplication" depends = "
            ...getFirstMatrix , getSecondMatrix"
            description = "Performs matrix multiplication when
                ...file1.txt and file2.txt are available">
            <grid-execute name = "Matrix Multiplication"
                environment = "PATH=/home/user1"
                directory = "/home/user1/MatMultProject"
                protocol = "GRAM"
                server = "node1.lisa.ecp.fr:2135"
                executable = "matrixMultiply"
                output = "file3.txt"
            />
        </target>
```

3.4.2.2   Grid Workflow Description Language

Grid Workflow Description Language (GWorkflowDL) [47] is a platform-neutral, XML-based workflow specification language based on the idea of high-level Petri-nets (HLPN). It incorporates the HLPN concept of edge expressions to model a service as a transition and conditions as a control flow mechanism. HLPN allows computation of output tokens of a transition based on its input tokens. Classical Petri-nets, however, allow only one type of token, which is generated every time a transition fires. Before starting the description of GWorkflowDL, we give a brief overview of Petri-nets.

Petri-nets are directed graphs. They consist of two sets of nodes: transitions (represented by rectangles or thick vertical lines) and places (represented by circles). Directed edges connect the places and transitions. An edge from place p to to a transition t is called an incoming edge for t and p is the *input place*. Similarly, an edge from a transition t to place p is called an outgoing edge and p is called the output place. Each place in the Petri-net can hold a certain number of tokens (called its *capacity*). Tokens represent the data flowing through the net. A transition is said to be enabled if there is a token present at each of its input places and all of its output places have not reached their capacity. Enabled transitions can fire by consuming one token from each of its input places and placing a token at each of its output places. Each edge (or transition) in a Petri-net can be assigned an edge expression. For incoming edges, variable names are used as edge expressions. Each transition may also have a set of boolean conditions. A transition may fire only when all the boolean conditions evaluate to true.

To explain the use of GWorkflowDL for describing grid workflows we consider the example of finding a solution to a system of linear equations using a matrix. A system of linear equations can be denoted by $AX = B$. The solution of the system of linear equations is given by $X = A^{-1}B$. It can be represented using a set of services as X=mult(inverse(A),B). Here mult is a service that takes two matrices as its parameters and finds their product. inverse finds the inverse of a matrix A defined as $A^{-1} = adj\,A/|A|$ where $adj\,A$ is the adjoint matrix of A and $|A|$ is its determinant.

Following is the HLPN representation of mult. Services are represented as transitions with its name written above the transition. Inputs and output are represented as places.

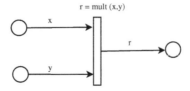

FIGURE 3.2: Petri-net representation of the mult service.

Application developers can also add *control places* to the graph. A control place has a simple token that does not have any value. So, input edges connecting a control place to a transition do not have any associated variables. Such edges are called *control edges*. Control edges are used for synchronizing the firing of transition with the corresponding control place. Consider the complete representation of our workflow as Petri-nets.

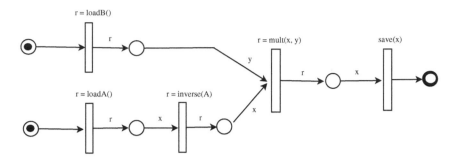

FIGURE 3.3: Petri-net representation for the solution of a linear system of equations.

Before the computation can begin, matrices A and B have to be loaded. These transitions do not require any input parameters. So we introduce a control edge for each of them. Control places contain one token that is consumed when the transition fires. Such transitions are executed only once.

A GWorkflowDL schema defines `<workflow>` as the root element. It may contain an optional element `<description>` having a human-readable form of the workflow being designed. `<workflow>` consists of several `<place>` and `<transition>`, which describe the Petri-net of the workflow. The `<transition>` element consists of platform-specific extensions such as `<WSRFExtension>` and `<JavaRMIExtension>`, which define the mapping of the transitions on specific grid platforms. `<transition>` also contains the `<inputPlace>` and `<outputPlace>` elements, which define the Petri-net. Following is an abbreviated GWorkflowDL representation of our `mult` service using the Java/RMI extension provided by GWorkflowDL.

```
<workflow>
    <place ID = "Matrix1"/>
    <place ID = "Matrix2"/>
    <place ID = "Result"/>
    <transition ID = "mult">
        <description> Performs matrix multiplication of
            ...two
                        compatible   matrices
        <description/>
        <inputPlace  placeID = "Matrix1"   edgeExpression
            ...= "a"/>
        <inputPlace  placeID = "Matrix2"   edgeExpression
            ...= "b"/>
        <outputPlace placeID = "Result"   edgeExpression =
            ... "result">
        <JavaRMIExtension >
            <method name = matrixMultiply.execute(x,y)/>
        </JavaRMIExtension >
```

```
</transition>
</workflow>
```

After selecting the appropriate services needed for an application, an application developer can create an abstract workflow using HLPN. The Petri-net then has to be adopted by assigning services and platform-specific edge expressions to transitions. As shown in the GWorkflowDL representation of `mult` method, a Java method of a remote interface providing the service is assigned to each transition and the variable names are assigned to input and output parameters. The resulting workflow can then be executed by the grid by assigning the appropriate host to each service.

3.4.2.3 Grid Services Flow Language

Grid Services Flow Language (GSFL) [48] is an XML-based workflow specification language, which allows describing grid services in the OGSA framework. It is described using XML schema like the GWorkflowDL. GSFL consists of the following important components:

- Service Providers: These are the services that take part in the workflow. The services are specified in the list of `serviceProviders`. Every service provider in the GSFL document has a unique name, specified as a part of the service definition. The service definition also contains the `type` of service provider. It specifies the type of the grid service as described by the WSDL document associated with the grid service. Services can be located statically or by looking up the registries using the `locator` element. Static lookup is performed by providing a static URL to an already running service. Services can also be started by *factories*. The handles for the factories are present in the GSFL document.

- Activity Model: The activity model lists all the activities in the workflow being described. These are the lists of operations provided by individual service providers. Each activity contains a `name` for identifying it. It also contains a `source`, which is the reference to a web service operation defined using `endPointType` element. The `endPointType` element contains the names of the operation, `portType`, `portName` and service name for a particular operation.

- Composition Model: It describes how new grid services can be created from existing ones. It describes the control and data flow between the services and also the direct communication between them. The composition model consists of two parts: Export model and Notification model. Export model contains a list of activities that need to be exported as operations of the workflow. A workflow instance composed from multiple grid services can also be viewed as a grid service. So, a workflow instance can be part of another workflow. This allows recursive definition of workflows. The data and control flow for each activity is described using `dataModel` and `controlModel` respectively. The dataModel describes the data flow when the exported activity is invoked. It contains two elements: `dataInTo`, which specifies the activity that receives the data provided as input to the exported activity, and `dataOutFrom`, which specifies the activity that would provide the output data to the exported activity. The controlModel consists of a chain of activities that need to be exported as operations of the workflow. The `controlIn` attribute of controlModel designates the first activity to be executed when the exported activity is invoked. Every controlModel also contains a series of controlLinks that defines the precedence list of activities to be invoked. The data contained in the dataModel and controlModel elements of GSFL document enable dynamic generation of the WSDL document and invocation of the exported activities. The second part of the composition model, the notification model, allows bypassing the workflow engine for carrying out every step of an activity. In web services workflow every time two services need to communicate, the data must pass through the web service workflow engine. When a large amount of data needs to be transferred this can be a potential bottleneck. In the grid services workflow model the data need not pass through the grid service workflow coordinator every time. The OGSA services communicate with each other using notificationSources and notificationSinks. Using `notificationLinks` the GSFL notificationModel links sources to links along with a *topic* of interest. If the services now need to transfer a large amount of data, they can do so without using the workflow engine as an intermediary.

- LifeCycle Model: The `lifeCycleModel` component of GSFL defines the order in which the services and the activities execute. The `serviceLifeCycleModel` contains a precedence link specifying the order in which services and activities are supposed to execute. The lifeCycleModel defines two types of scopes: *session* and *application*. In session scope no state is maintained between calls to workflow engine. A service is instantiated for each call using the serviceLifeCycleModel. In application scope a state is maintained between calls to the workflow engine. The services are instantiated only *once* for every instance of the workflow. This implies that not every call to the workflow engine is valid because the services implementing those activities may not be alive. The `activityLifeCycleModel` adds the constraint on some services so that they can be invoked only if certain other services have been invoked.

GSFL workflow is coordinated by a GSFL coordinator service that creates virtual ports and services allowing mapping to the processes internal to the workflow. We call the ports virtual because they do not physically exist on the GSFL coordinator. A GSFL workflow is started by creating a new instance of the GSFL coordinator using OGSA factory methods. The GSFL document pertaining to the workflow is then passed to this GSFL coordinator instance.

It generates a WSDL document containing all the newly exported operations. This WSDL can then be used by a client who wants to execute the workflow.

3.4.3   Workflow Scheduling Algorithms

We gave a brief idea about workflow scheduling in our discussion on Grid Workflow Management Systems. We discussed three workflow scheduling architectures: centralized, hierarchical and decentralized. In this section we focus on the algorithms that are used for workflow scheduling. Scheduling of jobs is an NP-Complete problem [49]. So, workflow scheduling employs some heuristic strategies for scheduling jobs that seek to optimize an optimization function. The optimization function consists of QoS parameters such as meeting a deadline or parameters such as idle-time of machines in the resource pool, throughput, communication time or overall job completion time. Scheduling decisions are also motivated by factors such as whether the application components have to be scheduled for single or multiple users and whether rescheduling is desired [50]. The performance of a scheduling algorithm is measured by the makespan of the task, which is the time taken to complete the entire workflow. Workflow scheduling strategies can be categorized as list scheduling algorithms, clustering algorithms, guided random search algorithms and task duplication based algorithms [51]. In list scheduling algorithms each task in the workflow is assigned a priority and tasks are selected for execution based on their priority. List algorithms produce a good quality schedule with less complexity. Examples include Levelized Min Time (LMT) and Heterogeneous Earliest Finish Time (HEFT). Clustering algorithms try to schedule those tasks on the same processors that exchange large amounts of data. *Task Duplication based Scheduling Scheme (TDS)* is an example of cluster based approach. Genetic Algorithms (GAs) are the most popular guided random scarch technique. GAs produce good quality schedule but their execution time is large. In task duplication algorithms, the same task is executed on more than one processor so that waiting time for dependent tasks is reduced. This approach has also been blended with list scheduling and clustering based techniques. Examples are *Critical Path Fast Duplication (CPFD)* and *Heterogeneous Critical Node First (HCNF)*.

We now present two workflow scheduling algorithms. The first algorithm uses a static approach to map application components of a workflow to grid resources. In the second algorithm an adaptive rescheduling strategy is used, which tries to enhance the performance of static scheduling algorithms. It is based on the Heterogeneous Earliest Finish Time (HEFT) algorithm, which is a list scheduling algorithm. Reference [50] presents the *Levelized Heuristic Based Scheduling (LHBS)* algorithm to schedule workflow applications onto grid. The paper argues that a better scheduling decision can be achieved by applying advance static scheduling to guarantee that the major computational steps in the workflow are executed on the best resources and the data movement is minimized. This is in contrast to the dynamic approach used by workflow management tools such as *Condor DAGMan* [43], which dynamically schedule workflow at every step. In the dynamic approach the scheduling decision is influenced by the state of the grid at those points when the application components are ready to execute. Now we describe the approach used by authors in [50] to schedule workflow applications. The workflow application is represented by a Directed Acyclic Graph (DAG). In the first stage of the algorithm a *rank* value is assigned to each resource on which a component can be mapped. The convention adopted for ranking is the lower the rank of a resource, the better suited it is to execute the application component. As a preliminary step of assigning ranks, it is checked whether the grid meets minimum application requirements such as OS type, memory size, cache size, etc. If a resource does not meet these requirements it is assigned a rank of infinity for that application component. For the eligible set of resources the rank value is obtained as a weighted linear combination of expected execution time on the resource and estimated data movement time. Let c_i be the application component and r_j be the resource whose rank value for c_i is being calculated. $map(c_i)$ is the resource on which c_i is mapped. $T(r_a, r_b)$ is the data transfer time between resources r_a and r_b. So the data movement time for resource r_j is calculated as the sum of $T(map(P), r_j)$ where P is the set of parents of c_i in the DAG. The estimated execution time is calculated by modeling the application behavior to different machine architectures. Machine architecture is defined by cache size, memory size and speed, number and type of execution units, etc. The performance model considers the application's memory access pattern and the number of floating point operations needed for different inputs of data.

After the rank has been assigned to resources for different application components, stage two of the algorithm begins. In this stage three heuristics are used to do the mapping between application components and resources. These heuristics are min-min, max-min and sufferage heuristic. All three heuristics are run and the heuristic that gives the minimum makespan is selected. e_{ij} is the expected completion time for component c_i on resource r_j. $mat(r_j)$ is the machine availability time of resource r_j. Then e_{ij} is defined as the sum of $mat(r_j)$ and $rank(i, j)$. $rank(i, j)$ is the rank of resource r_j for component $mat(c_i)$. Note that the expression for ECT used here is slightly different from the one used in Section 3.2. Here instead of expected execution time, rank is used, which encompasses both expected execution time and expected data transfer time. We can sum up the steps of the algorithm as follows:

1. First rank of the resources is computed for each application component.

2. After computation of ranks, the algorithm uses min-min, max-min and sufferage heuristic to find the best mapping. The mapping that gives the best makespan among the three is chosen as the final mapping.

Reference [64] presents a Heterogeneous Earliest Finish Time-based adaptive rescheduling *(AHEFT)* strategy for workflow applications in grids [51]. Theoretically, static scheduling algorithms provide an optimal scheduling pattern. However, this is a doubtful proposition in a dynamic grid environment. The performance of a job on a scheduled resource may change because of external or internal changes in the resource such as unavailability of resource due to network failure. Also, after static scheduling has been done, the grid resource pool can change. This happens due to availability of new resources, which were not available when the scheduling decision was made. Newly available resources may provide a better match for some of the application components thereby increasing the overall performance of workflow. AHEFT is based on the same logic. The AHEFT system architecture consists of two components: Planner and Executor. Planner consists of scheduler, performance history repository and a predictor. A scheduler instance is created for each workflow represented by a DAG. Planner finds resource availability by querying the grid information services. Scheduler makes its decision based on the data from the performance history repository and the predictor that finds communication and execution time for the available resources. The best mapping schedule is then submitted to the executor. While the application executes, the scheduler instance listens for pre-defined events such as resource pool change and sizeable variance in job performance on the scheduled resource. This may prompt the scheduler to reschedule the application if necessary. The scheduler instance also populates the performance history repository with the latest job performance. The executor of the AHEFT system consists of an execution manager, resource manager and performance monitor. The execution manager receives the schedule for DAG and executes it accordingly. It is also responsible for the staging of the input and output files. The resource manager allows advance reservation of resources per the execution schedule. The performance monitor informs the planner about job performance that can be used to make rescheduling decisions.

AHEFT uses a $n \times n$ data matrix of communication data where n is the number of jobs and $data_{i,j}$ is the amount of data that needs to be transmitted from job n_i to job n_j. A variable clock is used as a logical clock to measure the time span of workflow execution. After the successful completion of workflow, the clock contains its makespan. Let us assume a job n_i has been scheduled to run on resource r_j. Job n_i can start its execution only when it has the necessary data from all its predecessors. Further n_i cannot start before machine availability time, $mat(j)$, of resource r_j. The earliest start time (EST) of the job n_i on the resource r_j also depends on data transfer time for transferring output data from parents of n_i to the resource r_j. If n_i and its predecessors have been scheduled to execute on the same resource r_j, the data transfer time is zero. EST of a job n_i is the maximum of MAT and the maximum data transfer time for all the predecessors of n_i. The earliest finish time (EFT) of a job n_i on resource r_j is found by adding the expected execution time of job

n_i on resource r_j to the EST of n_i. Upward rank of a job n_i is defined as:

$$rank(n_i) = \overline{w_i} + \max_{n_j}(\overline{c_{i,j}} + rank(n_j))$$

where $\overline{w_i}$ is the average computational cost of job n_i and $\overline{c_{i,j}}$ is the average communication cost for data dependence of job n_j on n_i. For the exit job upward rank is defined as:

$$rank(n_{exit}) = \overline{w_{exit}}$$

The AHEFT algorithm starts by computing the ranks of all the jobs by traversing the graph upward from the exit job. The ranks are then sorted in decreasing order. Then the algorithm iterates over all the jobs in the sorted list. For each job in the list, every resource in the set of available resources is considered. The job is assigned to the resource that results in the minimum earliest finish time of the job. When a rescheduling relevant event (e.g., change in resource pool) occurs and the workflow is not complete, rescheduling of the remaining workflow may be made. The new list of resources is retrieved from the resource manager, which in turn obtains it from the grid information services. The performance estimation matrix, which contains the expected execution time for jobs on different resources, is updated based on the newly added resources. Then a new schedule is obtained by applying the AHEFT heuristic. If the makespan of the new schedule is less than the makespan of the older schedule, it is accepted, otherwise the old schedule is continued.

3.5   Fault Tolerance in Grids

A machine in a grid can break down due to various reasons or may be disconnected from the rest of the grid due to a network failure. It is important for grids to handle such cases in a manner transparent to the user. The reason for this is that a user visualizes the grid as one virtual machine. The user submits the application for execution and gets the result. The user should not be concerned with failures in the grid and should get the job done irrespective of it. So, failures should be handled by the grid without any user intervention. We can look at grid failures from two perspectives. A machine that was assigned to execute a job goes down. In this case the job has to be reassigned to a different machine that matches the user's criteria for the job and fulfills the QoS requirements, if any. In the other scenario, the job requested by the user may require data located at a remote location. But the network link to that location may be unavailable due to some reason. In this case, it is the responsibility of the grid to locate alternate data sources, using the grid information services. If alternate data sources are unavailable, the grid should generate an appropriate error message for the user. The two scenarios are dealt with in different manners. For the first case, the grid adopts a strategy known as *rescheduling*. We have already talked about rescheduling in workflow scheduling algorithms. The rescheduling we discussed in workflow scheduling dealt with assignment of a better resource to a job to minimize the makespan of the workflow application. Here rescheduling is concerned with fault tolerance in grids. Although the purpose is different, the approach is similar. Grid systems adopt a strategy known as *data replication* for the second scenario. In this the data is replicated at more than one site, so that there is an alternative data source if one of the sites goes down. Data replication is also done to enhance access locality as explained in Section 3.2.3. We have already discussed fault tolerance with respect to data replication in Chapter 2. So in this section we discuss fault tolerance with respect to the second aspect.

Reference [59] presents a classification of various types of faults that may occur in grids.

- *Hardware Faults*: Hardware failures due to a fault in components such as CPU, memory and storage devices. Accordingly, hardware faults can be classified as CPU faults resulting in incorrect output, memory faults due to defect in RAM, ROM or cache and storage faults, which may occur due to bad sectors.

- *Application and Operating System Faults*: These faults include memory leaks when an application does not release the memory allocated to it, operating system faults such as deadlock or improper resource management, unavailability of requested resource, etc.

- *Network Faults*: It happens as a result of physical damage to the networks. This may result in significant packet loss or packet corruption. Network faults also occur when individual nodes go down due to reasons like hardware faults.

- *Software Faults:* Such faults may occur due to unhandled exception condition such as division by zero or unexpected input when an incorrect file is supplied as the input to the program resulting in erroneous results.

Before we start a formal discussion of fault tolerance strategies, readers should keep in mind that a dynamic system such as a grid should be designed considering failures as a rule rather than exception.

3.5.1 Fault Tolerance Techniques

When a node assigned to perform a task or a part of the task (sub-task) fails, the task must be rescheduled to a different node. This can be viewed as three independent steps. The *first* step is the identification of failure. The grid meta-scheduler should detect a node failure. The *second* step is invocation of the grid information service to identify the available resources. The *third* and final step is mapping the task to the best possible resource available. It is not necessary that each of these steps be used by every fault tolerance mechanisms. Job replication mechanisms, for example, may not require a stringent failure detection criteria. Detection of failures is the most important aspect of fault tolerance. Reference [60] defines four important aspects of fault-tolerance in grids. These are:

Generic Mechanism for Failure Detection

Two kinds of failures need to be handled in grid applications: failure due to task crash and failure due to user-defined exceptions. Currently grid protocols such as GRAM do not have an in-built fault tolerance mechanism. Tools that use the GRAM protocol such as Condor-G and Nimrod-G adopt a custom-based fault tolerance approach that cannot be shared by other tools. The failure detection mechanism should not modify existing grid protocols. There are two reasons for this. The first is the backward compatibility problem: a system using an older protocol would not be able to communicate with the new system. The other problem is that local resource management systems would have to be modified to support the new grid protocol.

Flexible Failure Handling Strategies

A flexible failure handling strategy is required in grids to support multiple fault tolerance strategies. Every application should be given a chance to adopt a particular fault tolerance technique depending upon various available alternatives. The possible alternatives may depend of the type of application (mission-critical, long-running, etc.) or the nature of execution environment. Suppose a grid resource has a long down-time. In this case it would be more reasonable to start the application on another resource rather than restarting it on the same resource.

Separation Between Failure-Handling and Application-Specific Logic

The failure handling logic should be independent from the application-specific code. The reason for this is that an application may change over time. If the failure handling logic is hard-wired in the application-specific code, the code for handling failure needs to be rewritten every time the application is changed. Separation of concerns is important to make the grid application design simple. An analogy can be drawn from the TCP protocol where algorithms for handling failures and flow control are incorporated in the TCP layer and applications run over it. If we want to change the TCP layer, the application layer is not touched and vice-versa.

User-Defined Exception Handling

Users should be able to specify user-defined exceptions to handle task-specific failures based on the context of the task. The user should also be able to define custom procedures to handle these errors. Addition of user-defined exception handling in a fault tolerance framework gives the users the flexibility to define a course of action a task should take in case of failures, because a user can provide an intelligent approach to circumvent the failure based on the knowledge of the algorithm used in the task. Such an intelligent decision cannot be made by the grid scheduler.

Now we discuss some of the fault tolerance techniques described in grid literature.

3.5.1.1   Rescheduling

Reference [57] presents a rescheduling approach using stop and restart. The motivation for rescheduling is similar to the one presented in Section 3.4.3, rescheduling a job to a better resource. However, a similar strategy can be adopted for fault tolerance. In [57] when a running application is signaled to migrate, it *checkpoints* user-specified data. Checkpointing is the process of saving the state of the program including seek pointers for open files, so that the program can resume execution at a different machine. For fault-tolerance, instead of *user-directed* checkpointing such as the one in [57], an *automatic* checkpointing approach has to be followed. In automatic checkpointing, the state of the program is saved periodically to a persistent storage. When the machine running the program crashes, the program can be rescheduled to run on a different machine, continuing from the last checkpointed state. Checkpointing is critical in grid rescheduling because a grid application may take days to complete. If no checkpoint is maintained, the whole computation needs to be performed again, which is certainly not desirable.

3.5.1.2   Job Replication

Distributed Fault-Tolerant Scheduling (DFTS) [58] policy is based on the job replication mechanism. It does not reschedule the job when a fault occurs. Instead it schedules more than one copy of the job (called *replicas*) to a different set of resources. The job is considered complete if one of the set of resources successfully execute the complete job. The underlying assumption for job replication is that in a grid many resources may be under utilized. We now describe the system model adopted by [58] and describe the fault-tolerance mechanism with respect to it. The grid is assumed to have N sites under different administrative domains. Each of these can have different numbers of processors and storage machine. Every site contains a machine known as the *Single Resource Manager (SRM)*, which is responsible for scheduling and job staging services at that site. A single job can be allocated resources from multiple sites. Every SRM is assumed to be reachable from every other SRM in the grid unless there is network failure or the node hosting SRM goes down.

Each SRM maintains an Application Status Table (AST), which maintains state information for failure detection and recovery. DFTS assumes that the user provides sufficient information about the job, which can be used to estimate the runtime of the job using different candidate schedules. Every SRM is a backup of another SRM in the system. The primary SRM (PSRM) and backup SRM (BSRM) communicate periodically to exchange necessary information so that the BSRM can take over if the PSRM fails. If both PSRM and BSRM fail or cannot contact the remote SRMs, the SRM with the lowest id monitors the job execution.

When a job arrives at the home SRM (or PSRM), it looks for n candidate sites for job execution. n is assumed to be supplied by the user. If n sites are found, the job is scheduled on the available sites. If not, a set of sites equal to the number of unscheduled job replicas is *reserved* for future placement of the unscheduled replicas. When the site becomes available it contacts the SRM that reserved it. When the candidate sites for job replication are selected, one of the SRMs is designated by the home SRM (or PSRM) as BSRM. The other sites are informed about the BSRM by the home SRM. After this, the home SRM sends a replica of the job to each selected site and updates its job table. The job table information is also sent to the BSRM so that it can take over the PSRM in case the latter fails. Meanwhile, PSRM monitors the availability of the sites that have been reserved. As soon as one becomes available, the home SRM sends a job replica to it. A timer-based reservation scheme is used to avoid race cases. After an SRM offers its resources to execute a job, it does not accept a request from any other SRM until a time limit expires or it is released explicitly by the SRM which reserved it. The PSRM periodically queries the remote SRMs to find the status of the job replica running there. PSRM keeps a count of the number of *healthy* replicas and compares it to the pre-decided replica threshold. If the count falls below the threshold, a replacement is sought by polling all those sites where the job replica is not running. If a site is found, a job replica is forwarded to it and the job table and BSRM are updated. If no site is found, one of the sites is reserved. When the site becomes available, it contacts the SRM that reserved it. When a PSRM receives a job-complete status from any of the SRMs, it informs the other SRMs to suspend their job execution. The reserved SRMs are also released. The replication-based scheme is not suitable when the resources are scarce. In such cases a rescheduling-based approach for fault tolerance should be used.

3.5.1.3 Proactive Fault Tolerance

Reference [59] describes an agent-oriented proactive framework for fault tolerance in grids. Unlike rescheduling, which takes an action when a failure is detected, a *proactive* strategy monitors the grid for possibilities of failures and takes corrective measures before the actual failure might occur. Agents keep track of memory consumption of an executing process, hardware conditions, network resources and component mean time to failure (MTTF). Different agents keep track of different *types* of grid failures.

Hardware Fault Tolerance

Hardware Fault Tolerance is handled by two agents: *hardware monitoring agent (HMA)* and *hardware rebooting agent (HRA)*. HMA maintains information about previous pings, hardware MTTF and age. HMA also maintains historical information about node availability. The agent constructs a probability distribution based on this information and assigns score to each node. A new job is scheduled on the node with the highest score. HMA also monitors the nodes that are running a task. If the score of such a node falls below a threshold, the task is assigned to another machine with a better score. HRA reboots systems whose MTTF is very low even in the absence of load.

Application and Operating System Fault Tolerance

A *memory usage monitoring agent (MMA)* deals with the memory leaks in applications. It keeps track of total memory utilization during a specific time interval. A maximum limit on the amount of memory consumption is set initially. If the memory utilization for an application is high, MMA closely monitors the node. If the memory utilization is greater than the maximum memory limit for a substantial amount of time, the state of the application is saved and it is rescheduled to another node. A *micro rebooting agent (MRA)* monitors the processes running on a node and notices that a process is consuming exceptional amount of memory and has an unhealthy growth, it is killed and restarted. A system monitoring agent (SMA) keeps a record of uptime, operating system type, mean time between failure (MTBF), mean time to failure (MTTF) and age for every node. Like the HMA, SMA gives a score and rank to every node based on these criteria. Bonus and penalty is provided based on MTTF, MTTB, operating system name, age and uptime. If the score is below an acceptable threshold, the state of the job running on it is saved and the job is migrated to a healthy system. When scheduling new jobs, the unhealthy nodes are given less priority. The *resource tracker agent (RTA)* keeps a record of resources available on the system and whether they meet application needs or not. It also monitors the resources to consider possibility of deadlock. Elapsed time to complete an application and elapsed time of a scheduled application are also maintained. Based on these values, RTA checks for possible deadlock conditions or resource shortage. If the possibility is high, the job is rescheduled, otherwise the resource is reserved for the job.

Network Fault Tolerance

Two agents have been designed to counter network errors: *network monitoring agent (NMA)* and *node failure tracking agent (NFTA)*. NMA monitors the send-receive status of network packets to check whether the packets have been received properly or not. Further, it monitors how frequently a packet gets corrupted or lost. If packet loss or corruption is above a certain threshold, the job is migrated to another system. The NFTA keeps information about maximum network load, current network load, MTTF, and MTBF of individual nodes and decides about its state (healthy or unhealthy). It also performs load testing of the network and checks for malicious network attacks. Based on this information, NFTA may decide to migrate an application to another node.

All the agents interact with the grid scheduler and grid information services and may reschedule a job proactively based on the conditions we have discussed.

3.5.2   A Framework for Fault Tolerance in Grids

Reference [60] describes a two-phase approach to fault tolerance in grids. The *first* phase is the failure detection, which is handled by a *generic failure detection service*. Failure recovery is the *second* phase handled by a *flexible failure handling framework*. The generic failure detection service is based on an event notification mechanism. The failure handling framework uses a workflow-structure to handle task-level and workflow-level failures. The techniques to handle task-level failures mask the task crash without affecting the workflow execution. The techniques to handle workflow-level failures allow changes to the flow of workflow execution. These techniques deal with user-defined exceptions and task crash failures that cannot be handled by the task-level failure handling technique. Workflow-level failure handling allows users to specify corrective measures based on their knowledge of the workflow context. Task-level techniques include retrying, checkpointing and replication. Workflow-level techniques include the concept of workflow redundancy, for example, by launching several instances of the same task using different algorithms.

The failure detection service consists of two components: notification generators and notification listeners. Notification generators include grid generic server (GRAM), which generates *event* notification messages such as *task active*, *task failed*; heartbeat monitor, which generates periodic *aliveness* notification messages on behalf of its monitoring task and the task itself, which generates *task-specific* event notification messages such as *file not found exception detected*. Based on these messages the notification listener can detect whether the task has completed successfully or not. It can also distinguish between a task-crash failure and a user-defined exception. Tasks can register to the heartbeat monitor service by calling an event notification API inside its code. The state of a task is determined by combination of messages from the three sources. For example a combination of a *Done* message from the grid server and an *End Task* message from the task implies that a task has completed successfully. Now suppose the task crashed due to some unhandled exception. In this case the *Done* message is received but the *End Task* message is not received from the task. The grid client can conclude that the task did not complete successfully.

The failure handling framework defines an *XML workflow process definition language (XML WPDL)*, which allows a user to define workflow as a directed acyclic graph (DAG). It mentions the action to be taken when a failure is detected. Task-level failure handling includes retrying, replication and checkpointing. Retrying can be thought of as rescheduling on the same machine without checkpointed data. A strategy for job replication is discussed in the next section. Checkpointing is not a stand-alone fault tolerance technique. It is coupled with either retrying (starting the job on the same resource) or rescheduling (resuming execution of the job from its checkpointed state at a different node). The following fragment of XML WPDL shows how retrying is specified.

```
<Activity name = 'PartialDifferentiation' max_tries = '5'
  interval = '15'>
    <Input>...</Input>
    <Output>...<>Output>
</Activity>
```

Replication is specified by defining `policy='replica'` in the task definition as shown below.

```
<Activity name = 'PartialDifferentiation' policy =
'replica'>
    <Input>...</Input>
    <Output>...</Output>
    <Implement> Differentiation </Implement>
</Activity>
...
<Program name = 'sum'>
    <Option hostname = 'node1.lisa.ecp.fr'../>
    <Option hostname = 'node2.lisa.ecp.fr'../>
    <Option hostname = 'node3.lisa.ecp.fr'../>
    <Option hostname = 'node4.lisa.ecp.fr'../>
</Program>
```

For the above XML WPDL, the task is simultaneously submitted to four hosts in the grid. Workflow-level failure handling techniques include alternative task execution, workflow-level redundancy and user-defined exception handling. In alternative task execution, when a task fails to execute an alternative task is executed. The alternative task may implement a different algorithm than the original task to perform the same function. In workflow-level redundancy different tasks are executed for the same computation in parallel. This technique is useful in situations where tasks having different execution characteristics are available for the same computation. Suppose task A and task B take the same input and produce the same output but are implemented differently. Both these tasks are executed in parallel and even if one of them succeeds, the workflow can proceed normally. User-defined exception gives specific treatment to a task failure. An example of such a technique could be alternative task execution.

References

[27] Ian Foster and Carl Kesselman. *The grid: blueprint for a new computing infrastructure*, chapter 2, pages 15–52. Series in Computer Architecture and Design. Morgan Kaufmann, 2nd edition, December 2003.

[28] Howard Jay Siegel and Shoukat Ali. Techniques for mapping tasks to machines in heterogeneous computing systems. *Journal of Systems Architecture*, 46(8):627–639, 2000. Available online at: `citeseer.ist.psu.edu/siegel00techniques.html` (accessed January 1st, 2009).

[29] David Fernandez-Baca. Allocating modules to processors in a distributed system. *IEEE Transactions on Software Engineering*, 15(11):1427–1436, November 1989. Available online at: `http://portal.acm.org/citation.cfm?id=76150` (accessed January 1st, 2009).

[30] Muthucumaru Maheswaran, Tracy D. Braun, and Howard J. Siegel. Heterogeneous distributed computing. In J.G. Webster, editor, *Encyclopedia of Electrical and Electronics Engineering*, volume 8, pages 679–690. John Wiley & Sons, New York, NY, 1999. Available online at: `citeseer.ist.psu.edu/maheswaran99heterogeneous.html` (accessed January 1st, 2009).

[31] Tracy D. Braun, Howard Jay Siegel, Noah Beck, Ladislau Bni, Muthucumaru Maheswaran, Albert I. Reuther, James P. Robertson, Mitchell D. Theys, Bin Yao, Debra A. Hensgen, and Richard F. Freund. A comparison study of static mapping heuristics for a class of metatasks on heterogeneous computing systems. In *Heterogeneous Computing Workshop*, volume 8, pages 15–29. IEEE Computer Society, 1999. Available online at: `citeseer.ist.psu.edu/braun99comparison.html` (accessed January 1st, 2009).

[32] Muthucumaru Maheswaran, Shoukat Ali, Howard Jay Siegel, Debra A. Hensgen, and Richard F. Freund. Dynamic matching and scheduling of a class of independent tasks onto heterogeneous computing systems. In *Proceedings of the Heterogeneous Computing Workshop*, volume 8, pages 30–44. IEEE Computer Society, 1999. Available online at: `citeseer.ist.psu.edu/maheswaran99dynamic.html` (accessed January 1st, 2009).

[33] M. Srinivas and Lalit M. Patnaik. Genetic algorithms: a survey. *Computer*, 27(6):17–26, June 1994. Available online at: `http://dx.doi.org/10.1109/2.294849` (accesed January 1st, 2009).

[34] Robert Armstrong, Debra Hensgen, and Taylor Kidd. The relative performance of various mapping algorithms is independent of sizable variances in run-time predictions. In *HCW'98: Proceedings of the 7th Het-*

erogeneous Computing Workshop, pages 79–87. IEEE Computer Society, 1998.

[35] Steven Fitzgerald. Grid information services for distributed resource sharing. In *HPDC'01: Proceedings of the 10th IEEE International Symposium on High Performance Distributed Computing*, pages 181–194. IEEE Computer Society, 2001.

[36] Laszewski, Warren Smith, and Steven Tuecke. A directory service for configuring high-performance distributed computations. In *HPDC'97: Proceedings of the 6th IEEE International Symposium on High Performance Distributed Computing*, pages 365–375. IEEE Computer Society, 1997.

[37] Rob Byrom, Brian Coghlan, and Andrew Cooke. R-GMA: a relational grid information and monitoring system. Web Published, 2003. Available online at: `http://www.r-gma.org/pub/Cracow-2003-rgma.pdf` (accessed January 1st, 2009).

[38] Brian Tierney, Ruth Aydt, Dan Gunter, Warren Smith, Valerie Taylor, Rich Wolski, and Martin Swany. A grid monitoring architecture. Technical Report GWD-PERF-16-2, Global Grid Forum, January 2002. Available online at: `citeseer.ist.psu.edu/tierney02grid.html` (accessed January 1st, 2009).

[39] Giovanni Aloisio, Massimo Cafaro, Italo Epicoco, Sandro Fiore, Daniele Lezzi, Maria Mirto, and Silvia Mocavero. iGrid: a novel grid information service. *Advances in Grid Computing*, 3470:506–515, July 2005. Available online at: `http://www.gridlab.org/WorkPackages/wp-10/Documents/igrid-egc.pdf` (accesed January 1st, 2009).

[40] Globus Alliance. A Globus primer. Web Published, 2007. Available online at: `http://www.globus.org/toolkit/docs/4.0/key/GT4_Primer_0.6.pdf` (accessed January 1st, 2009).

[41] Jia Yu and Rajkumar Buyya. A taxonomy of workflow management systems for grid computing. Technical Report GRIDS-TR-2005-1, University of Melbourne, Grid Computing and Distributed Systems Laboratory, Australia, March 2005. Available online at: `citeseer.ist.psu.edu/yu05taxonomy.html` (accessed January 1st, 2009).

[42] Soonwook Hwang and Carl Kesselman. Grid workflow: a flexible failure handling framework for the grid. In *HPDC'03: 12th IEEE International Symposium on High Performance Distributed Computing*, pages 126–137. IEEE Computer Society, June 2003.

[43] Condor. DAGMan. Web Published, 2007. Available online at: `http://www.cs.wisc.edu/condor/dagman/` (accessed January 1st, 2009).

[44] Condor. What is Condor ? Web Published, 2007. Available online at: http://www.cs.wisc.edu/condor/ (accessed January 1st, 2009).

[45] Gregor von Laszewski, Beulah Alunkal, Kaizar Amin, Shawon Hampton, and Sandeep Nijsure. GridAnt: client side grid workflow management with Ant. Web Published, 2007. Available online at: http://www.globus.org/cog/projects/gridant/gridant-whitepaper.pdf (accessed January 1st, 2009).

[46] Gregor von Laszewski, Kaizar Amin, Mihael Hategan, Nestor J. Zaluzec, Shawon Hampton, and Sandeep Nijsure. GridAnt: a client-controllable grid workflow system. Web Published, 2007. Available online at: http://www-unix.mcs.anl.gov/~laszewsk/papers/vonLaszewski--gridant-hics.pdf (accessed January 1st, 2009).

[47] Martin Alt, Sergei Gorglatch, Andreas Hoheisel, and Hans-Werner Paul. A grid workflow language using high-level Petri-nets. Technical report, CoreGRID, January 2006. Available online at: http://citeseer.ist.psu.edu/alt06grid.html (accessed January 1st, 2009).

[48] Sriram Krishnan, Patrick Wagstrom, and Gregor von Laszewski. GSFL: a workflow framework for grid services. Web Published, 2002. Available online at: http://citeseer.ist.psu.edu/krishnan02gsfl.html (accessed January 1st, 2009).

[49] Micheal R. Garey and David S. Johnson. *Computers and intractability: a guide to the theory of NP-completeness.* Series of Books in the Mathematical Sciences. W.H. Freeman & Co, 1979.

[50] Anirban Mandal, Ken Kennedy, Charles Koelbel, Gabriel Marin, John Mellor-Crummey, Bo Liu, and Lennart Johnsson. Scheduling strategies for mapping application workflows onto the grid. In *HPDC'05: 14th International Symposium on High-Performance Distributed Computing*, pages 125–134. IEEE Press, July 2005.

[51] Haluk Topcuouglu, Salim Hariri, and Min you Wu. Performance-effective and low-complexity task scheduling for heterogeneous computing. *IEEE Transaction on Parallel Distributed System*, 13(3):260–274, 2002. Available online at: http://dx.doi.org/10.1109/71.993206 (accessed January 1st, 2009).

[52] Ramin Yahyapour. *Design and evaluation of job scheduling strategies for grid computing.* PhD thesis, University of Dortmund, November 2002. Available online at: https://eldorado.uni-dortmund.de/handle/2003/2838 (accessed January 1st, 2009).

[53] David A. Lifka. The ANL/IBM SP scheduling system. In *Proceedings of the Workshop on Job Scheduling Strategies for Parallel Processing*, volume 949 of *Lecture Notes in Computer Science*, pages 295–303, London UK, 1995. Springer-Verlag.

[54] Srividya Srinivasan, Rajkumar Kettimuthu, Vijay Subramani, and P. Sadayappan. Selective reservation strategies for backfill job scheduling. In *JSSPP'02: 8th International Workshop on Job Scheduling Strategies for Parallel Processing*, pages 55–71. Springer-Verlag, 2002.

[55] Uwe Schwiegelshohn and Ramin Yahyapour. The NRW metacomputing initiative. In *Workshop on Wide Area Networks and High Performance Computing*, pages 269–282. Springer-Verlag, 1999.

[56] Yang Zhang, Charles Koelbel, and Ken Kennedy. Relative performance of scheduling algorithms in grid environments. In *CCGRID'07: Proceedings of the 7th IEEE International Symposium on Cluster Computing and the Grid*, pages 521–528. IEEE Computer Society, 2007. Available online at: http://dx.doi.org/10.1109/CCGRID.2007.94 (accessed January 1st, 2009).

[57] F. Berman, H. Casanova, and A. Chien. New grid scheduling and rescheduling methods in the GrADS project. *International Journal of Parallel Programming*, 33(2–3):209–229, June 2005. Available online at: http://dx.doi.org/10.1007/s10766-005-3584-4 (accessed January 1st, 2009).

[58] Jemal H. Abawajy. Fault-tolerant scheduling policy for grid computing systems. In *Proceedings of 18th International Parallel and Distributed Processing Symposium*, volume 14, page 238, April 2004.

[59] Mohammad Tanvir Huda, Heinz W. Schmidt, and Ian D. Peake. An agent oriented proactive fault-tolerant framework for grid computing. In *E-SCIENCE'05: Proceedings of the 1st International Conference on e-Science and Grid Computing*, pages 304–311, Washington, DC, USA, 2005. IEEE Computer Society. Available online at: http://dx.doi.org/10.1109/E-SCIENCE.2005.15 (accessed January 1st, 2009).

[60] Soonwook Hwang and Carl Kesselman. A flexible framework for fault tolerance in the grid. *Journal of Grid Computing*, 1(3):251–272, September 2003. Available online at: http://www.springerlink.com/content/x71543229j403x62/ (accessed January 1st, 2009).

[61] Srikumar Venugopal, Rajkumar Buyya, and Lyle Winton. A grid service broker for scheduling distributed data-oriented applications on global grids. In *MGC'04: Proceedings of the 2nd Workshop on Middleware for Grid Computing*, pages 75–80. ACM Press, 2004. Available

online at: `http://doi.acm.org/10.1145/1028493.1028506` (accessed January 1st, 2009).

[62] Henri Casanova, Arnaud Legrand, Dimitri Zagorodnov, and Francine Berman. Heuristics for scheduling parameter sweep applications in grid environments. In *Proceedings of 31st Heterogeneous Computing Workshop*, pages 349–363, May 2000. Available online at: `citeseer.ist.psu.edu/casanova00heuristics.html` (accessed January 1st, 2009).

[63] Kavitha Ranganathan and Ian Foster. Decoupling computation and data scheduling in distributed data-intensive applications. In *HPDC'02: Proceedings of the 11th IEEE International Symposium on High Performance Distributed Computing*, page 352. IEEE Computer Society, 2002.

[64] Zhifeng Yu and Weisong Shi. An adaptive rescheduling strategy for grid workflow applications. In *IPDPS'07: IEEE International Parallel and Distributed Processing Symposium*, pages 1–8, March 2007.

Chapter 4

Security in Grid Computing

4.1 Introduction

Security in grids differs from the Internet security due to the challenges that arise when we seek to build scalable virtual organizations (VOs). As discussed in Chapter 1, VOs are a group of geographically distributed individuals or organizations having permanent or temporary existence created to share resources and services among themselves. This sharing is governed by a set of rules or policies that define the conditions and extent for that sharing. The dynamic and cross-organizational nature of the virtual organization makes the problem of implementing security in grids a challenging one. The problem is complicated by the fact that there is no central point of control in grids. Absence of a central point of control in grids means that each resource provider in the grid has to make an assessment of the risk before interacting with any other service provider. To understand the process of risk assessment and subsequent trust establishment, we need to know the traditional security areas that play a vital role in defining security in the grids. These include authentication, authorization and confidentiality. You might already be familiar with these terms. However, here we explain these terms keeping in mind the grid security requirements.

4.1.1 Authentication

Authentication establishes the identity of an entity in the network. This entity may be a user, a process or a resource. Suppose you want to communicate with another computer on the network that claims to have a particular identity. The process of authentication enables you to validate that claim. The simplest authentication means can be the use of a username and a password. Another means of network authentication is Kerberos, which provides authentication to client/server applications by using symmetric key cryptography. The technology that plays the pivotal role in authentication in grids is Public Key Infrastructure (PKI). PKI describes the security system used to identify entities through the use of X.509 certificates. These identity certificates are issued by highly trusted organizations known as certifying authorities (CAs). A user may play different roles or may be part of different virtual organizations at the same time. Different VOs might have agreed upon use of different CAs. So users in grids may use different credentials at the same point of time.

4.1.2 Authorization

Authorization is the second step of trust establishment between two entities in the grid. It refers to the verification of the privileges assigned to an entity to access a resource or service provided by the grid. Authorization is done only after authentication has been performed successfully. In grids, the resource owners should have the ability to grant or deny access to a resource, based on the identity or membership of a virtual organization, of an entity.

One of the earliest authorization techniques used in grids was the Globus Toolkit Gridmap file. This file contains a list of distinguished global names of grid users and the corresponding local account names to which they are mapped. The function of access control is performed by the host operating system along with the local security policies. This method entails a lot of effort on the part of the local system administrator in maintaining an updated version of the gridmap file. The Community Authorization Service (CAS) [65] takes the ease and manageability of grid authorization to the next level. In CAS a resource owner can grant restricted access to a resource owned by it to the virtual organization. Then it is up to the community to allocate these resources to those who need them. Each VO consists of a CAS server, which acts as a trusted intermediary between the resources and the users of the VO. Virtual Organization Membership Service (VOMS) [66], developed by the European Union DataGrid projects, provides for the management of authorization information within VOs spanning multiple organizations. VOMS provide a database containing user roles and capabilities and a set of tools to access and manipulate the database. The tools can use the database contents to create credentials when a user needs them. These credentials contain the basic authentication information that the standard grid proxy certificates contain. In addition to that, they also contain the role and capability information from the VOMS database running on a VOMS server. VOMS-aware applications can make use of the VOMS data present in these credentials to make authorization decisions. These credentials are also compatible with the standard grid applications, which do not use the VOMS data. Such applications can make use of the credentials without the VOMS data present within.

4.1.3 Confidentiality

Confidentiality means hiding sensitive information from those who do not have the rights to access them. Grid applications often involve number crunching on sensitive data such as industrial information, financial information, medical data, etc. The need for protecting such data may arise either due to intellectual property rights issues or due to their privacy. The most basic approach to confidentiality followed in cryptography is data encryption. Grids store sensitive information in the online database. Access to such data must be carefully planned and controlled. Grid applications often require running critical code on a remote resource. The remote resource may be either a stand-alone machine or a dynamically chosen cluster according to the decision made by the load balancer in the grid. Such selection must contain only those nodes in the cluster that are trusted by the data owner.

So we see that grid applications have some specific security access requirements, different from Internet security. This entails development of grid security architecture with the specific grid security perspective. Grid security is built upon well-known security standards. In this chapter we will talk about the existing security technologies. After having an understanding of these technologies we talk about the application of these standards to grid security. This will provide a clear understanding of the working of the encryption techniques within the grid security framework.

4.2   Trust and Security in a Grid Environment

Resource sharing is fundamental to the grid definition. The initial use of grids was to support scientific collaborations, where the interacting entities knew each other. Such collaborations have an implicit trust relation as all participating entities have a common objective and hence it is assumed that resource sharing and its access would be free from any malicious intent. However, with the growing popularity of grids, it is widely being used in the business domain where a resource might be shared with unknown third parties. Such situations involve a certain amount of risk as the resource owner cannot distinguish between genuine users and users with malicious intent. So grid users must have a way to assert their identity.

Trust is the fundamental aspect of any secure communication. Various definitions for trust can be found in the literature [67]. Security in traditional systems focus on providing security from unsolicited users by providing access to only authorized users. The situation is slightly different in grid environments. Grids provide a mechanism for discovering resources to the users in grids. A malicious user can, however, provide incorrect or misleading information. Authors in [68] talk about how trust-based systems can be used as protective measures in such a situation. In [69] a classification of various trust-based systems has been made. It defines trust as a relationship between truster, the subject that trusts the target entity, and trustee, the entity that is

trusted. It talks about a formal approach for trust establishment, evaluation
and analysis for Internet-based services. This article discusses trust in context
of distributed systems and hence can form the basis of understanding trust in
grid systems.

Resource sharing can occur in grids only when the two parties involved in
the sharing are able to assert each other's identity. As in any other secure data
exchange, this is done in grids using authentication followed by authorization.
However, unlike other distributed systems, grids have some specific security
requirements [70]. We will discuss these requirements and their solutions in
GSI and OGSA security.

4.2.1   Existing Security Technologies

In this section we cover those security technologies that have been successfully deployed in various existing security systems. All of these technologies
are based on open standards and form an integral part of grid security. Of
these technologies, Kerberos is not explicitly the part of existing grid security
architecture, but can be used as an authentication mechanism to provide security in client/server architecture. As we shall see, Kerberos also provides
some of the functionalities desirable in grids like single sign-on and delegation
of privileges using Ticket Granting Ticket (TGT). The same functionality is
provided by X.509 proxy certificate, which is a part of the Grid Security Infrastructure (GSI). However, the creation and delegation of Kerberos TGTs
require involvement of a trusted third party (KDC). On the other hand an
X.509 proxy certificate can be created without the involvement of a third
party.

We start our discussion with Public Key Infrastructure (PKI), which forms
an integral part of GSI. In PKI we talk about the X.509 digital certificates,
which form an integral part of PKI. Next in this section, we cover the Kerberos
network authentication protocol explaining the key components of Kerberos
and the steps involved in the authentication mechanism. In the end we discuss
GSI, the existing security infrastructure used in the grids today.

4.2.1.1   Public Key Infrastructure

Public Key Infrastructure (PKI) provides users a way to do secure communication in insecure public network using public/private key pair. PKI
involves a trusted third party, which is called a certifying authority (CA).
The CA issues a digital certificate to users who may be an individual or an
organization. The digital certificate uniquely identifies a user. The digital
certificate follows the structure specified by the X.509 system. In an X.509
system a distinct name for the user of the certificate is bound with its public key by a CA. The private key of the certificate is securely kept with the

owner of the certificate while the digital certificate containing the public key is available for public use. A piece of data signed by the private key can be decrypted only using the public key and vice-versa. This forms the basis of the PKI.

PKI follows a hierarchical structure for establishing a chain of trust. At the lowest level of the hierarchy are the end users who are issued with the digital certificate. At the next level are the CAs who are authorized to issue certificates on a regional level. There is no fixed specification for the size of a region. It may be as small as an organization or as big as a state or a country. Each of these CAs also has a digital certificate. These certificates are in turn signed by another CA, which is at a higher level in the hierarchy, and this may continue. At the topmost level are those CAs that issue certificate for smaller CAs. It's important to note that there can be more than one CA at the top of the hierarchy. These CAs are in the business of issuing digital certificates and are trusted by everyone. Suppose a user obtains the digital certificate of an individual. The user then examines whether the certificate has been signed by a trusted CA. If it does not trust the CA, which has signed the certificate, it may request the digital certificate of the CA, which is in turn signed by another CA, at an upper layer in the hierarchy. This may continue until the user finds that the certificate has been signed by a CA which it trusts. This chain of trust is important because a user may trust very few recognized CAs. The hierarchical structure of PKI is illustrated in Figure 4.1. Although we have shown only three levels in the diagram for simplicity, in practice there can be more than three levels.

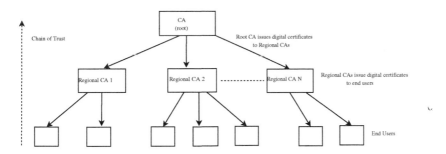

FIGURE 4.1: Hierarchical structure of Public Key Infrastructure (PKI).

X.509 digital certificate contains the following fields [71]:

1. **Version** - It specifies the version of the X.509 digital certificate. If extensions are used, the version is 3. If extensions are not used but a

`UniqueIdentifier` is present, the version is 2. If only the basic fields are used the version is 1.

2. `SerialNumber` - Every CA must assign a unique serial number to a certificate it issues. So CA name and the serial number uniquely identify any certificate.

3. `AlgorithmIdentifier` - This identifies the algorithms used in the certificate. For instance `md5withRSAEncryprion` indicates that RSA is used for the generation of the public/private key pair and `md5` is used as the hashing algorithm.

4. `Issuer` - This field contains the name of the CA that has signed the certificate.

5. `Validity` - The validity field consists of two fields: *"not valid before"* and *"not valid after"* specifying the duration for which the certificate is valid.

6. `Subject` - It contains the name of the organization or individual that owns the certificate.

7. `SubjectPublicKeyInfo` - This field is composed of two fields: *"algorithm"*, which specifies the algorithm used to generate the *"public/private key pair"*, and *"subject public key"*, which contains the public key of the certificate holder.

8. `IssuerUniqueId` - This is an optional field specifying a unique ID for the issuer of the certificate (CA).

9. `SubjectUniqueId` - This is again an optional field specifying a unique ID for the owner of the certificate.

10. `Extensions` - It is an optional field that may contain extension to the basic fields.

11. `CertificateSignature` - Signature consists of the hash of the entire certificate (except the certificate signature field), signed by the private key of the CA. So to verify the genuineness of a certificate a user may compute the hash of the entire field using the algorithm specified in the certificate. Then decrypt the hash signed by the CA using the CA's public key. If these two values of hash are the same, the certificate has not been tampered with and can be trusted.

    Because of the nature of the X.509 digital certificate, it can be transmitted over an insecure communication channel. This is because if any of the fields is modified, the hash of the modified certificate and the hash present in the certificate will not match and hence the certificate can be considered invalid. X.509 digital certificates are used in the Secure

Socket Layer (SSL), a transport layer security protocol. As we shall see in the subsequent sections, X.509 digital certificates form the basis of authentication mechanisms in GSI.

The CAs in the PKI are also responsible for publishing the Certificate Revocation List (CRL). This list contains the serial number of those certificates that have become invalid because their validity has ended or because of some fraudulent activity. A certificate may also be revoked if the private key corresponding to the certificate has been compromised. Such a list can be published on the CA's website or can be maintained in an online repository. The CRLs are also signed by the CA that issues them. This is done to prevent generation of false or incorrect CRLs.

4.2.1.2 Kerberos

Kerberos is a network authentication protocol developed by MIT. It is a distributed authentication protocol that provides mutual authentication to client and server using symmetric-key cryptography. Symmetric-key cryptography means that the same key is used for both encryption and decryption of the message.

We explain some terms now used frequently in the Kerberos literature. These terms will be widely used in the subsequent sections. So the readers are advised to understand these terms carefully before proceeding further.

- Principal - Principal is the entity whose identity is being verified.

- Verifier - Verifier is the entity that verifies the identity of the principal.

- Kerberos Realm - A realm is the administrative domain in which Kerberos functions. Kerberos realms usually consist of the Internet Domain Name of the organization written as uppercase letters.

- Security Principals - Entities recognized by the Kerberos realm. It may either be a user or a process running on behalf of the user.

The Kerberos protocol is partly based on the Needham and Schroeder authentication protocol [72]. However, some changes have been incorporated in its design to make it more secure and suitable for client-server authentication. These changes include use of timestamps to reduce the number of messages required for the initial authentication. A ticket-granting service has been added for ease of use. This facilitates subsequent authentication without re-entering the principal's password. Support has been provided for *cross-realm authentication* (authentication of the principal in a different Kerberos realm than the verifier). We will discuss cross-realm authentication later in this section.

Kerberos does not send any critical information across the network that may be used by an attacker to impersonate the principal or the verifier. Any

Kerberos authentication process involves three parties: client, server and a trusted intermediary, referred to in the Kerberos protocol as the Key Distribution Center (KDC).

Key Components of Kerberos Authentication System

1. *Key Distribution Center* - KDC is the trusted third party. It runs as a service on a physically secure system. It maintains a database containing account information for all the security principals in its realm. The KDC is divided into two entities: Authentication Server (AS) and Ticket Granting Server (TGS).

2. *Authentication Server* - It shares a long-term secret key with all the entities in the Kerberos realm, namely the security principals and the Ticket Granting Server (TGS). This long term secret key is often called the *master key*. The master key is generated using a hash function from the security principals' password. The purpose of AS is to issue the Ticket Granting Ticket (TGT), which is used by the client to authenticate to the Ticket Granting Server (TGS).

3. *Ticket Granting Server* - Ticket Granting Server or TGS is a trusted third party that uses short-term keys known as Ticket Granting Ticket (TGT) to provide tickets to clients that want to authenticate to the server. All the security principals in the Kerberos realm share a TGT with the TGS. TGT is generated by the AS using the master key the security principals share with the AS. After the client has successfully authenticated with the TGS, the TGS issues a ticket that is used by the client to access the desired server.

4. *Ticket* - It is used in Kerberos by the client to authenticate to the server. It contains the information needed for the authentication. The ticket contains both plain text and encrypted text. The encrypted text contains information intended for the server. Two kinds of tickets are used in Kerberos: Ticket Granting Ticket (TGT) issued by the AS to the client and the normal ticket issued by the TGS to the client.

Now we explain the Kerberos authentication protocol in detail. The Kerberos protocol can be understood as a series of three steps. The first step involves message exchange between the client and the AS. The client and AS use the master key for their mutual authentication. At the end of this step a TGT is issued to the client to communicate with the TGS. The TGT contains the authentication information required by the client to prove its identity to the TGS. In the second step the client presents the TGS issued by the AS to the TGT. If a client wants to communicate with more than one server, the second step can be repeated provided the TGT issued to the client does not expire. The outcome of this step is a ticket to be used by the client in the third step of communication with the server. As with the second step, the third step can be repeated a number of times between the same set of client and server without execution of steps one and two. The only condition is that the ticket must be valid. In this stage, the client can also seek authentication from the server, if a mutual authentication is required. At the end of the third step, the client is granted access to the requested resource.

The client and server can use the session key used during the authentication process to protect the integrity and privacy of communication. The Kerberos system defines two types of messages that can be used during communication, namely *safe messages*, which can be sent over the network as plain text, and *private messages*, which are sent as encrypted text to ensure the privacy of the message. However, the parties involved in communication can use the message type that suits their particular application needs.

We now describe the stages of Kerberos authentication protocol with reference to the message exchanged between the communicating parties [73].

Step I: Client and the Authentication Server message exchange

1. *Kerberos Authentication Service Request (KRB_AS_REQ)* - The client sends a request to the authentication server of the KDC in plain text using its registered identity. To prove its identity, the client puts pre-authentication data, which consists of the timestamp encrypted using the client's master key in the message. The master key is generated by applying an appropriate hash function to the client password. The client can request the AS to grant a TGT valid for a specific duration. A random number called *nonce*, which has not been used earlier, is also sent in the KRB_AS_REQ. This value is used to prevent attempts of replay attacks.

2. *Kerberos Authentication Service Reply (KRB_AS_REP)* - The AS, upon receiving the KRB_AS_REQ message, looks for the client in its database and extracts the associated master key. It then decrypts the pre-authentication data and verifies the timestamp inside. If the timestamp is acceptable, AS can assume that the data has been encrypted using the client's master key. Next the AS generates a random key to be used as the session key for the client server communication. The KRB_AS_REP message contains the Ticket Granting Ticket (TGT) for the client to be presented to the Ticket-Granting Server and a session key. The entire message is encrypted using the client's master key. TGT contains the client's authorization data and the session key encrypted using the symmetric key the TGS shares with the AS. The client does not know its master key. When the KRB_AS_REP message arrives, the user types his or her password. The password and appropriate hash function is used to generate the master key, if the password is correct. The password is destroyed immediately after the generation of the master key. The client can now use the master key to extract the session key and the ticket from

the KRB_AS_REP message. If an error occurs during the authentication process, a KRB_ERROR message is sent by the AS specifying the error code.

3. When the client receives the KRB_AS_REP, it extracts the session key and the TGT. Both TGT and the session key are stored in a cache at the client. This concludes the step I of the Kerberos authentication protocol.

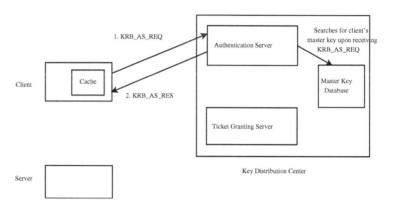

FIGURE 4.2: Client and authentication server message exchange (Kerberos step I).

Step II: Client and Ticket Granting Server message exchange

This exchange is initiated by the client and occurs when it wants to obtain authentication credentials for a target server or when it wants to validate or renew an existing TGT. The messages in this set of communication are encrypted using the session key obtained in the first step.

1. *Kerberos Ticket Granting Service Request (KRB_TGS_REQ)* - It is sent by the client to the TGS. The KRB_TGS_REQ message includes the TGT acquired from the AS server, identity of the target server whose credentials are being requested and an authenticator message similar to one sent by the client in step I to AS, encrypted using the user's session key.

2. *Kerberos Ticket Granting Service Response (KRB_TGS_RES)* - Upon receiving the KRB_TGS_REQ message, the TGS decrypts the TGT using its master key. It then extracts the client's session key from the TGT. The session key is used to decrypt the authenticator message

from the client. If the authenticator is acceptable, the TGS generates a new session key for to be shared between the client and the server. A KRB_TGS_RES message, which contains two copies of the newly generated session key, is sent to the client. One copy is encrypted using the client's session key. The second copy is contained in a ticket along with the client's authorization data, encrypted using the target server's master key.

3. The client decrypts the target server session key and stores it in its cache along with the ticket obtained from the TGS.

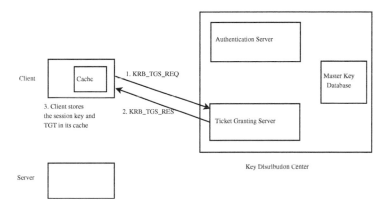

FIGURE 4.3: Client and ticket granting server message exchange (Kerberos step II).

Step III: Client Server Message Exchange

At this point the client is ready to initiate an authentication process with the server.

1. *Kerberos Application Request (KRB_AP_REQ)* - This is the first message in the authenticated transaction between the client and the server. This message contains the ticket for the communication with the server encrypted using the server's master key, an authenticator encrypted using the session key for the server and a flag indicating whether the client seeks mutual authentication or not.

2. *Kerberos Application Response (KRB_AP_RES)* - The server decrypts the ticket using its master key. It then extracts the client's authorization data and the session key for client server communication. Using

this session key the server decrypts the authenticator and evaluates the correctness of the timestamp for its acceptance. After this, the server looks for the mutual authentication flag in the KRB_AP_REQ. If the flag is set, the server encrypts the authenticator message using the session key shared with the client. This message is sent to the client with a message type KRB_AP_RES.

3. The client decrypts the message using the session key. The decrypted result is matched to the authenticator message sent by the client to the server in KRB_AP_REQ. If both strings are identical, the client can assume that it is communicating with the desired server and the connection can proceed.

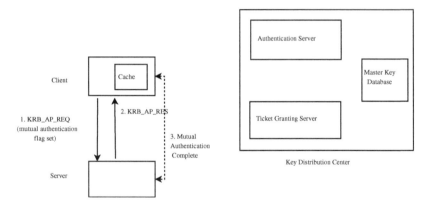

FIGURE 4.4: Client server message exchange (Kerberos step III).

Cross-Realm Authentication in Kerberos

We saw in the previous section how a client authenticates to a server in the same Kerberos realm. However, we can configure a Kerberos system such that security principal in one realm can communicate with security principal in another realm. This is called cross-realm authentication.

KDCs in different realms share an inter-realm secret key. Any two realms sharing an inter-realm secret key can trust each other. When a client wants to talk to a server in another realm, it sends a request to the TGS in the same realm. The TGS realizes that the server is in another realm and so a ticket cannot be issued. Instead of issuing the service ticket, the TGS in the client's realm issues a TGS referral. The TGS referral is the KRB_TGS_REQ message. It contains information to indicate that KRB_TGB_REQ refers to a TGS in another realm. It also includes a TGT for the referred TGS. Using this TGT the client creates a KRB_TGS_REQ for the TGS in the server's realm. The TGS in the server's realm can use its inter-realm secret key to decrypt the referral ticket. If the decryption is successful, it sends the client a service ticket for the intended server in its realm. Cross-realm authentication is shown in figure 4.5.

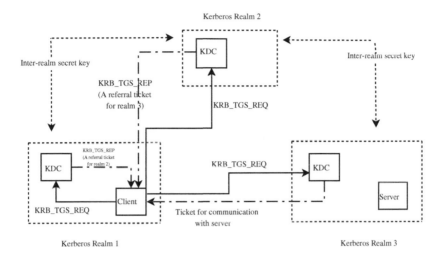

FIGURE 4.5: Cross-realm authentication in Kerberos.

Kerberos realms have a hierarchical organization wherein each realm shares a distinct key with each of its child realms and a key for its parent realm. When the number of realms increases, it becomes infeasible for each KDC pair to maintain an inter-realm secret key. So the Kerberos realms have been designed to have a hierarchical organization wherein each Kerberos realm shares a key with its parent and a distinct key with each of its child. Cross-realm authentication, however, suffers with some shortcomings:

1. It is vulnerable to denial-of-service attacks where the attacker can be in different Kerberos realms.

2. If an intermediary KDC is unavailable, the authentication cannot proceed.

3. The client must perform TGS exchange with each KDC of the trust path. If the number of KDCs on the trust path increase, there is a significant communication overhead.

4. The clocks of the KDCs of the target Kerberos realm and client Kerberos realm must be synchronized for authentication to succeed.

Advantages of Kerberos Authentication Protocol

We list the advantages of Kerberos over other authentication schemes.

1. As the passwords are not sent over the network in any form (plain text or cipher text), Kerberos is not susceptible to eavesdropping.

2. Kerberos supports single sign-on which allows users to authenticate just once to the AS and request service tickets for different servers from TGS using the same valid TGT.

3. Kerberos supports mutual authentication, which ensures that both parties know that they are talking to the genuine target and not an imposter.

4. Message exchange uses timestamp information to prevent replay attacks. Also, in Kerberos, all the tickets have an expiration time. This time is usually small enough to minimize the chances of replay attack.

5. The session key used between the client and server can be used to encrypt the messages providing integrity and privacy of messages.

6. Kerberos is based on open Internet standards and a wide variety of implementations is available.

7. Kerberos supports authentication delegation, which means that a service can access local or remote resources on behalf of the user.

Disadvantages of Kerberos Authentication Protocol

1. The security of the entire Kerberos system depends on how secure the KDC is. If the security of KDC is compromised the security of the entire system is compromised.

2. Kerberos relies on security principal's password as the source of its identity. So if the password is stolen or guessed by the attacker, he/she can impersonate the actual user. So it is advisable to choose a strong password and refrain from keeping the password file on local disk.

3. As Kerberos uses mutual authentication for client and server, it is important for applications running on both the parties to be Kerberos aware. Many legacy systems have to be rewritten to support Kerberos authentication.

4. In Kerberos the authenticator messages are valid for a certain duration. If an attacker captures such packets on the network, it has a certain time window to use the same services as the client who generated the authenticator messages.

4.2.1.3 Grid Security Infrastructure

Grid Security Infrastructure (GSI) is part of the Globus Toolkit. As the name suggests it defines the complete architecture that provides the necessary functionalities for the implementation of security in grids. GSI has been developed to meet some of the specific security needs of grids. These are:

1. Single Sign-On - A user should be able to authenticate itself once and then access the grid resources without requiring any further authentication. A proxy certificate is created the first time the user authenticates. The proxy certificate can be used to repeatedly authenticate for a period mentioned in the duration of the proxy certificate.

2. Delegation of privileges - Consider a situation where there are three entities X, Y and Z. Y trusts X and Z trusts X. Now suppose X wants Y to perform a task that requires access to a resource on Z. But as Z does not trust Y, it cannot allow Y to access its resource. To solve this problem, X can delegate a part of its privileges to Y by issuing a proxy certificate. The format of proxy certificates and how they are generated will be explained in detail in the section on proxy certificates. The proxy certificate along with its private key is called as the proxy credential. The proxy credential provides Y restricted access to those resources on Z that are needed to complete X's task. Restricted access means that, although X might be authorized to access other resources on Z, access is provided through proxy certificates to only those resources that are needed to complete X's task. The delegation of privileges in this way is called credential delegation. Consider another scenario where a user has submitted a job to a grid resource. Now the user wants to delegate a subset of its privilege to the job to access other resources on the user's behalf. This situation is different from the first scenario. In the first scenario, the *user* is provided the proxy credential and in the second scenario only the *job* is given to the proxy credential.

3. Inter-domain security support - A grid can contain entities spanning multiple organizations, each of which may follow a different intra-domain security policy. A grid security solution must be able to provide support for interaction among entities located in different security domains. However, access to a local resource is governed by local security policies decided by the local system administrator. So, there must be a proxy or an agent running on the local system on behalf of the remote client to provide access to the local resource.

4. Secure Communication - There may be a need for secure message exchange among entities in the grid. We shall see GSI provides such support using transport layer security (TLS).

5. Authentication & Authorization - A grid security solution must provide for secure and scalable authentication and authorization. In the subsequent sections we will describe how GSI caters to these needs.

6. Uniform credentials - As grids require inter-domain interaction, there must be a uniform way to represent credentials of the entities in the grid. GSI uses X.509 certificate format for representing the credentials of the entities in the grid.

To support these requirements, GSI provides various functions, namely, authentication, authorization, delegation and message protection. Each of these tasks is handled by an open standard. These form part of the GSI.

X.509 End User Certificates

X.509 certificates are used to identify entities in the grid. Each user in the grid is identified by a unique X.509 certificate and a process running on behalf of that user makes use of the same certificate. The format of the X.509 certificate was described in Section 4.2.1.1 on PKI. GSI also supports the basic username- and password-based authentication along with WS-Security. However, when such authentication is used, features such as delegation, single sign-on and integrity are not available.

SAML and Globus GridMap File

The Globus gridmap file contains a list of global names that are authorized to access a service on that node. All authorized users are also mapped to the local username. This file has to be updated manually. In a production environment having a large number of users, the gridmap file does not scale well. As described in [74], all users may have a different set of permissions and this may require frequent updating of the gridmap file.

Due to the limitations of the gridmap file found in earlier versions of GT, GSI in GT4 uses SAML assertions for making authorization decisions. SAML stands for Security Assertion Markup language. It is an XML standard for exchanging identity, authentication and authorization data between two entities. SAML describes the format of security *assertions* that can be used by service provider to make authorization decisions about the identity provider. SAML assertion consists of one of the three statement types, namely, authentication statement, attribute statement or authorization decision statements [75]. SAML assertions are specified in XML using XML schema. SAML generally uses SOAP protocol binding, which means that SAML assertions are transmitted as SOAP messages. HTTP is used as the communication protocol. SAML assertions can be embedded in X.509 certificate chain to provide authorization information to the service provider.

Suppose a user has been authenticated by a service provider through the user's X.509 end entity certificate. The service provider then sends an *AttributeQuery* to the user. This SOAP message contains encrypted DistinguishedName from the users X.509 end entity certificate. The service provider signs the attribute request so that its origin and integrity can be verified. The user then sends a *response* message containing the required attributes. The response is signed by the user so that the service provider can verify that it has come from a genuine source. If the service provider is able to establish that its a genuine response, it provides the user an access to the desired resource [76].

X.509 Proxy Certificates

Proxy certificates are issued by an entity for delegating a subset of privileges owned by it to another entity. A duration is specified in the proxy certificate stating the duration for which it is valid. As the proxy certificate is valid for a limited duration, the private key of the proxy certificate can be kept on the local filesystem without any encryption. The operating system file access policies can be used to prevent easy access to the private key. A proxy consists of a *new* public/private key pair. The new certificate contains the owner's identity and contains a field to indicate that it is a proxy certificate. This certificate is signed by the owner instead of a CA.

When proxy certificates are used, the authenticating party receives the proxy certificate and the owner's X.509 certificate. The owner's public key is used to test the validity of the proxy certificate (whether it has been signed by the owner or not). Similarly the CA's digital signature present in the owner's X.509 certificate can be used to validate the owner's certificate. In this way a chain of trust is established from the CA to the owner and from the owner to the proxy.

Message Protection

Message protection can be done at two levels: at the transport level using *transport layer security (TLS)* or at the message level using the *WS-Security* standard and the *WS-SecureConversation* specification. TLS provides for both privacy and integrity of data. If TLS is used, the entire communication is encrypted. On the other hand, message level security encrypts only the content of the SOAP message which means that the format of the SOAP message, is preserved. As both transport-level and message-level protection use X.509 certificates, they can provide services such as integrity and privacy. These services can be enabled or disabled according to the application needs. Transport-level security is the default message protection technique in the Globus Toolkit 4 [77].

4.2.2   Emerging Security Technologies

Now we move on to some emerging security standards. First we cover the *Web Services Security (WS-Security)* and the other web service specifications that provide an extension to WS-Security. Then we discuss Open Grid Services Architecture (OGSA), an emerging service-oriented architecture for grids. OGSA, being a service-oriented architecture, makes use of the WS-Security standard and its extensions. In OGSA, first we cover the fundamentals of OGSA, to give our readers an understanding of OGSA. Then we describe the OGSA security model, explaining the features OGSA seeks to add to the grid security.

4.2.2.1   Web Services Security

Web Services Security (WS-Security) is the standard to provide security features such as integrity, privacy, confidentiality and single message authentication to SOAP messages. The SOAP messages are sent over HTTP. So the technologies that secure web applications like *Secure Socket Layer (SSL)* and basic username- and password-based authentication can be used to secure the web services. However, these techniques secure the entire data sent over the wire and not just the SOAP messages. So to obtain security at the message level rather than at the transport level WS-Security was developed. This has an added implication. The SOAP messages can be secured irrespective of the underlying transport protocol using WS-Security. So in short, WS-Security provides an extension to the SOAP messages to build secure web services. WS-Security has been designed to work with a wide variety of security protocols including PKI, Kerberos and SSL. According to the WS-Security specification [78], *"WS-Security provides support for multiple security tokens, multiple trust domains, multiple signature format and multiple encryption technologies"*.

The support for multiple technologies adds flexibility to WS-Security. It helps in negotiating and finalizing a technology that can be used by both the entities involved in the communication. WS-Security has been developed with the goal to make it as flexible as possible. To make this possible, the WS-Security specification does not impose any constraint on the security protocols used. It gives the liberty to choose any set of protocol, which can be used for secure SOAP message exchange.

WS-Security adds security to a SOAP message using the following three levels of security:

1. XML Encryption - This enables encryption of the SOAP message body or a part of it to ensure confidentiality of the message. Algorithms that can be used for block encryption include triple DES, AES-128, AES-192 and AES-256. Message digests require SHA-1 and recommends SHA-256. Algorithms for stream encryption, key transport, key agreement

and message authentication can be found at [79].

2. XML Digital Signatures - This enables SOAP messages to be digitally signed ensuring the integrity of the message. XML digital signatures specify the recommended algorithms and their respective URIs in the XML document. Algorithms have been specified for message digest, encoding, message authentication code (MAC), and signature. The details of the XML digital signature specification can be accessed at [80].

3. Authentication tokens - WS-Security specification defines the various token types that can be embedded in the SOAP message header. Security tokens are used for single-message authentication. They represent a set of claims made by the client, which may include name, identity, key, certificate, privilege, etc [78]. To keep the design of WS-Security extensible, no specific token type has been defined. However, it does mention the supported token types. The WS-Security specification mentions support for following token types:

   (a) Username Security Tokens - It consists of a username and optional password information.

   (b) Binary Security Tokens - WS-Security draft specifies the encoding for two binary security tokens. These include X.509 certificates and Kerberos tickets. The WS-Security draft also describes the syntax to be used in the SOAP message header. The details can be found in the WS-Security specification document.

These security methods can be used in combination to provide fine-grained security to the application. Other web service specifications are also associated with WS-Security architecture. These include:

1. *WS-SecureConversation* - It is a web service extension built on top of WS-Security and WS-Trust. WS-Security stresses authentication of one message but does not provide a security context. WS-SecureConversation provides for the security of more than one message.

2. *WS-Federation* - It is a specification, for standardizing how organizations share user identities in heterogeneous authentication and authorization systems. It also provides for brokering of trust and security token exchange. Optional features include the hiding of identity information of the entities. The complete specification can be found at [81].

3. *WS-Policy* - The WS-Policy framework provides a specification for the service requestor and service provider to enumerate their capabilities, needs and preferences in the form of *policies*. The details about the specification can be found at [82].

4. *WS-Trust* - The web services trust language is an extension to the WS-Security specification. It defines additional constructs and primitives for request and issue of security tokens and ways to establish the presence of trust relationships with parties in a different trust domain. The complete specification can be found at [83].

5. *WS-Privacy* - This specification has only been proposed and is not yet published. It will use a combination of WS-Security, WS-Policy and WS-Trust for communicating privacy policies among organizations. The organizations that deploy web services, and expect the incoming SOAP request to contain an assurance of compliance with the service provider's privacy policies, should use this service. In short, WS-Privacy will allow the interacting parties to agree over a privacy policy, confirming to the requirements of both the parties.

4.2.2.2   OGSA Security

Introduction to OGSA

OGSA stands for Open Grid Services Architecture. OGSA defines a grid architecture based on web services. OGSA has been designed as a combination of grid systems and web services to provide solution to these challenges. Its goal is to standardize the grid services like resource management, security management, and job management by providing standard interface for these services. The OGSA is built using two technologies: the Globus Toolkit and web services. The components of the Globus Toolkit that are relevant to the discussion of OGSA are the grid resource allocation and management (GRAM) protocol, which provides support for remote job submission and control, GRAM's gatekeeper service providing secure and reliable service creation and management, Meta Directory Service (MDS) for resource discovery and Grid Security Infrastructure (GSI) for providing authentication, authorization, message integrity and delegation services [84].

When we talk about grids, the focus is on resources like storage resources, network, databases, computational resources, etc. However, when we talk about OGSA, the stress is on services. All the resources we just talked about are modeled as services. The entities using these services need to know the interface definition of these services. Then, they can invoke these services like any other web service through exchange of messages. For the application developers the task becomes simpler, because now they have to work with standard service interfaces. The main idea of service oriented architecture is to provide a common interface for different implementations. OGSA introduces a new term, *grid service*. Grid service is a web service with extensions to make it suitable for grid-based applications. Grid services add a lot of functionalities to the web services. A web service is *stateless*, which means that a web service doesn't remember what the user did in his last request. Another problem with web services is that they are persistent. This means, their life-time is strictly bound to the web service container. This is not desirable as the next client who invokes the web service could easily access the information from the previous client. To solve this, grid service maintains a factory for a web service, which is responsible for the creation of web service instances. As grid services are created dynamically, a Grid Service Handle (GSH) is used to distinguish one grid service instance from another. Instance-specific information such as protocol binding, method definition, and network address are maintained in Grid Service Reference (GSR). Unlike GSH, whose value does not change once created, the value of a GSR can change during its lifetime. Grid service provides support for life-cycle management, which is absent in web services. Grid service adds another important feature absent in web service called service data. Service data allows us to add structured data (classes, array, etc.) to the WSDL interface. A grid service can be configured to provide notification to the clients about change in grid services to which they subscribe. Grid services add the notion of service groups to web services. It allows different services to be grouped together and accessed through a single entry point.

Web service definition language (WSDL) is used for the definition of the service interfaces independent of the underlying protocol used for service invocation. A single service interface may be bound to multiple protocols. For example an interface may be bound to HTTP or local *Inter Process Communication (IPC)*. Multiple protocol binding is important because the service requestor and service provider may be on the same machine. In such cases the overhead of network protocols is undesirable. On the other hand to support cases where service provider and requestor are on different machines, network protocol binding to the service is mandatory. It's important to mention here that the implementations of the service interfaces providing the same functionality may vary. This stems from the fact that these services may be running on different operating systems that provide different native services.

In a grid environment, clients using the grid service must be aware of any change in the service interface. They should also know whether the newer version of the service is compatible with the older version. The clients should also be aware of the protocol binding of the new service definition to enable them to choose the correct network protocol to be used for communication with the service provider. OGSA provides this functionality. It lets the client update its knowledge about the service definition, for example, to find the operations supported in the newer service interface and the protocol bindings used by the newer service. OGSA also provides support for discovery of available services, dynamically create new services, provide a mechanism by which a service may inform other services about its changing state.

OGSA provides a collection of standard interfaces that define the set of operations they support. Grid services implement one or more of these interfaces to provide the corresponding operations supported by these interfaces. For example some operations supported by the `GridService` interface are: `FindServiceData` to find different information about a grid service instance, `SetTerminationTime` to set the time when a grid service instance terminates and *Destroy* to terminate a grid service instance and free the allocated resources. A custom grid service can extend the `GridService` interface to provide user-defined functions. For example a grid service named `BasicmathService` may implement functions for addition, subtraction, multiplication and division in addition to the ones provided by the `GridService` interface. All the grid services must implement the `GridService` interface. List of other interfaces defined by OGSA can be found in [84].

OGSA Security Model

The introduction to the OGSA section should have given the readers a good idea about how and where OGSA fits in the grid scenario. Now we move on to how security issues concerning grids are addressed by OGSA. OGSA security seeks to address the grid security issues such as identity mapping, authentication, authorization, confidentiality, integrity, diverse VO policies, secure credential delegation, etc. The problem with grid security is that the VOs may span multiple organizations and it's not easy to formulate security policy for such cross-organizational scenario. The situation gets further complicated by the fact that an organization may be part of different VOs at the same time. OGSA security model sets out to provide a solution to the problem of grid security in a platform-independent and language-neutral way.

An OGSA security architecture should cover the following prime aspects of security [85]:

1. Authentication - To provide an authentication mechanism independent of the underlying technology used for authentication.

2. Delegation - To make provision for delegation of a subset of complete access privileges to a requestor for the invocation of a service.

3. Single Sign-on - A user should need to authenticate only once to make use of an OGSA service. Subsequent access to that service should not require re-authentication for a certain period of time.

4. Credential Lifespan and Renewal - In situations where the job submitted by a user would take longer to complete than the lifespan of the initial delegated credential, the user should be notified before the expiration of the credential so that the credential may be renewed for the completion of the job.

5. Authorization - Allowing access to OGSA services to only those who are authorized to do so.

6. Confidentiality - Confidentiality here is the same as described in the beginning of this chapter with a slight modification. Confidentiality here is applicable for a SOAP message, the building block of OGSA-based communication.

7. Integrity - Integrity of a message means that the receiver receives the original copy of the message and not a copy modified during transit. So, there must be a way to find whether the message has been modified during transit.

8. Logging - Logging refers to maintaining a log containing a timestamp for any events that occur within any of the services. Such information could be used for auditing and non-repudiation.

9. Firewall Traversal - Firewalls form an important barrier as far as cross-organizational barriers are concerned. So, the OGSA security model should provide a mechanism for firewall traversal without modifying or compromising the local security policy.

Keeping these requirements in mind we move on with the description of the OGSA security model. The model must ensure that whenever a grid service is accessed, local policy constraints of the system that host the grid service are met. OGSA security makes use of the existing web services security standards. It's important to understand that the OGSA security model does not make use of any particular security protocol. It is rather a collection of various security protocols. For example security at the network layer can be achieved using IPSec; at the transport layer SSL/TLS can be used to implement security. For message-based security XML standards like XML Encryption, XML Digital Signature and SAML can be used. So similar security services can be provided at different layers. As OGSA is a service-oriented architecture, it makes extensive use of the WS-Security standards. As described in the section on WS-Security, the WS-Security architecture is a collection of various modules. These include WS-Security, WS-Policy, WS-Federation, WS-Authorization, WS-Policy, WS-SecureConversation and WS-Security. The OGSA security specification is based on specification of these modules. If a particular web service security specification does not meet the requirements of OGSA security model, those specifications should be modified accordingly to provide easy integration into the OGSA security model [85].

Now we talk about the security specifications of OGSA as provided by the Global Grid Forum (GGF) [85]. At the time of this writing the OGSA specification was not complete. So there may be some changes to the OGSA specification in the future. OGSA security specification covers the following aspects:

1. OGSA places importance on *naming* of users, services and groups. This is necessary for authorization through policy evaluation and enforcement. For a correct policy evaluation, a unique name should be provided to the entities across different domains. Further logs are generated for finding the usage pattern and auditing in the grid. These logs may be referred to by any entity in the grid. So a uniform naming convention should be followed.

2. An entity may need to access a remote service that may have different security policies and mechanisms. The remote service provider may be part of a different organization and may use a different credential format. So, OGSA service must support seamless conversion of identity between different domains and also the conversion of credential format.

3. OGSA should provide *support for multiple authentication mechanisms* such as PKI and Kerberos. This is desirable because different organizations may use different authentication mechanisms.

4. In a grid scenario a message may pass through several intermediaries. Transport layer security provides end-to-end security. So when intermediaries are involved, end-to-end security cannot be established. For such a scenario OGSA should provide session-based security over a series of message exchanges, belonging to the same security context. This requirement is addressed by WS-SecureConversation. However, at the time of this writing, the WS-SecureConversation specification is in progress. So for the time being Generic Secure Services API (GSSAPI) can provide the session-based security.

5. When two parties communicate in a grid, one of the parties requests a service, and the other provides it. Before such a request is made, it is necessary that the local authorization policy is enforced at the requester, i.e., whether the requester is allowed to access a particular service at the provider. The provider must enforce a policy check to make sure that it can serve the request. So, policy enforcement is done at both ends. There may already be existing policy enforcement tools in an organization. So OGSA must provide support for a pluggable authorization module to ease the deployment issues.

6. The OGSA specification addresses various security policies management issues. These include authorization policy management, trust policy management and privacy policy management. These policy enforcements make use of WS-Security extensions such as WS-Trust, WS-Privacy, WS-Policy and WS-Authorization.

7. As you must be aware by now, a VO may span over multiple organizations and may consist of a large number of organizations. So the developers of the OGSA specification devised a set of services to help

VO policy information be consistently distributed across the members of the VO. The architecture defines a set of OGSA services from which VO membership and policy information can be retrieved.

8. It is necessary in grids for entities to have the flexibility of delegating a subset of rights to another entity that can carry out the specified task on its behalf. Grids already support delegation of credentials using proxy certificates. However, the delegation mechanism envisioned in the OGSA should be independent of the underlying authentication mechanism. Delegation in the OGSA specification makes use of SAML, Extensible rights Markup Language (XrML), Extensible Access Control Markup Language (XACML) and WS-Security specifications.

9. OGSA services should traverse firewalls and network address translating (NAT) routers just like the web traffic. As OGSA is a service-oriented architecture and needs to send messages to users informing about an important event, it's important that such messages reach the intended recipient. As in some of the earlier specifications, the *OGSA firewall interoperability specification* makes use of WS specifications such as WS-Routing, WS-Policy and WS-Referral.

10. In a service requester/service provider scenario it is imperative that the service requester is aware of the security policies of the service that it wants to access. This necessitates a way to exchange the security policies among the two parties. The OGSA specification makes use of the WS-Policy component of the WS-Security extensions for *security policy exchange*.

11. The OGSA specification makes provision for logging all the authentication and authorization data. The *OGSA audit service specification* provides a management interface and the policy for logging, for example, maintaining filters on the types of logs that can be stored in the system.

4.3 Getting Started with GSI

As we have discussed, GSI is used for implementing the security functionalities in Globus Toolkit 4. It is based on a set of open standards such as X.509 certificates, TLS, etc. We have already covered the theoretical aspects of GSI. In this section we will talk about the functional aspects of GSI. We will see how X.509 end user certificates can be acquired from *SimpleCA*, how credentials are managed and how proxy certificates are created and acquired.

4.3.1   Getting a Certificate

Under normal circumstances you would need a X.509 certificate from a certifying authority (CA). However, if you are deploying a grid for testing purposes, you can use the SimpleCA package that comes bundled with the Globus Toolkit 4 (GT4). It provides a wrapper around the OpenSSL CA functionality and can be used for testing purposes in simple grid examples. It should, however, be used only when a CA is not available.

To start, you should install the GT4 available freely from the Globus website [86]. For the subsequent discussions in this section, we will assume that GLOBUS_HOME is the Globus installation directory. For help on installing GT4 and setting up user accounts with necessary privileges, refer to Appendix A. We now show the steps of configuring a SimpleCA present in GT4.

1. A setup script is present to setup a new SimpleCA. This script needs to be run once per grid. Execute the script as:

   ```
   GLOBUS_HOME / setup / globus / setup - simple - ca
   ```

2. This will ask you to enter the necessary information about the SimpleCA. This includes:

   (a) common name (cn) - This identifies the certificate of the SimpleCA as the CA certificate. CA certificate is the certificate of the CA of the grid and is used to verify the signature on the certificates issued by the CA.

   (b) organizational unit (ou) - This is your fully qualified domain name (FQDN). This name differentiates between the CAs of different Globus domains created using the SimpleCA script.

   (c) organization (o) - It identifies your grid.

3. Then you will have to configure the CA's email and expiration date.

4. Next you will be prompted to enter a password for the SimpleCA. This will be used when signing the certificate by the CA.

   This completes the configuration steps of SimpleCA. To complete the configuration for GSI, you need to run the following script:

   ```
   GLOBUS_HOME / setup / globus_simple_ca_CA_Hash_setup / setup - gsi
   ... -default
   ```

After the completion of this step, you can create the host certificate. Once a host certificate is created, users can send request for signing the certificate to the SimpleCA. To create a certificate for a user, run grid-cert-request as the user. This prompts for a pass-phrase and after it has been entered, three files are created in the .globus directory present in the home directory of the user. These files are `usercert.pem` (empty), `userkey.pem` and `usercert\_request.pem`. The `usercert\_request.pem` file has to be sent to the SimpleCA at the email address to which it was configured. The SimpleCA then signs the certificate using the `grid-ca-sign` script. The signed certificate is then mailed back to the user. This certificate is then placed in the same directory where it was generated, and is renamed as `usercert.pem` thereby replacing the empty file. This certificate can now be used for authenticating the user to access various grid resources and services.

4.3.2 Managing Credentials

MyProxy, an open source software, is used for managing user X.509 certificates. A grid user may use different machines to access the grid services. So the user must have the grid credentials on every machine. However, this process has associated security threats as this requires manually copying the certificates to every machine. MyProxy provides a clean solution for such scenarios. MyProxy maintains an online repository to manage user certificates. This way the long-term private key of the user is not transmitted over the network. MyProxy can also be used by users to delegate proxy credentials to those services that access a grid resource on the user's behalf. So MyProxy can be used to store both X.509 end-entity certificates and proxy credentials. It also provides support for renewing user credentials in situations where the lifespan of the delegated credential is not sufficient for the completion of the job submitted by the user. Users can store credentials from multiple CA on the same MyProxy server.

To store the credential in the online MyProxy credential repository, use the `myproxy-init` command. This should be done on a computer that stores your credentials. The `myproxy-init` command prompts you for the passphrase you entered while creating you grid credentials using the command `grid-cert-request`. Then you are prompted for a new password used to secure your credentials in the MyProxy repository and needed when you access your credentials. You can specify the MyProxy server on which you want your credentials to be stored, if there are more than one MyProxy servers. `myproxy-init` uses the files userkey.pem and usercert.pem present in the .globus directory, which is in turn present in the user's home directory (refer to the Section 4.3.1, Getting a Certificate). This proxy credential can be retrieved whenever needed using the `myproxy-logon` command. This command prompts you for the password you entered while creating the proxy credentials using `myproxy-init`.

We now consider a situation where there are three sites A, B and C. Site A contains the X.509 end entity certificate. Site C runs a grid service (say a GridFtp server). Now site A wants site B to access the GridFtp server on site C, on its behalf. In this scenario, site A can delegate a proxy credential to a MyProxy server and inform site B about this. Site B can then retrieve the proxy credential from the MyProxy server and access the grid service. Before starting the actual communication with the MyProxy server, the client establishes a secure, authenticated channel with the MyProxy server. Once the TLS handshake is complete, the client sends one byte with zero value to the server. The server must ignore this byte. After this, the actual MyProxy protocol [87] starts. We explain the MyProxy message sequence using the figure shown below. Please note that a credential is a collection of certificate and the private key.

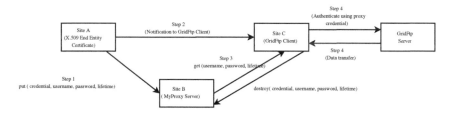

FIGURE 4.6: Using MyProxy server for delegation of credentials and subsequent access to GridFtp server.

1. Site A runs a MyProxy java client. It generates a random username and password for the proxy credential to be delegated to the MyProxy server. The java client calls the put() method passing it its credential, username, password and lifetime as parameters. The following steps occur during a put request.

   (a) The MyProxy server generates a public/private key pair and sends a certification request [88] to the java client running at site A. A certification request consists of the newly generated public key, a distinguished name and some optional attributes. The whole thing is signed by the MyProxy server.

   (b) This certificate request is signed by the private key from the X.509 end entity certificate of site A. This signed public key with some other fields (described in the previous section on proxy) forms the proxy certificate.

   (c) The MyProxy server stores the credential on a local filesystem. A standard response indicating failure or success is returned to site A.

2. Notification - Site A informs site B about the delegated proxy credential and the location of the MyProxy server. It provides the username, password and the lifetime parameters to site B as well.

3. Site B now has all the necessary parameters to retrieve the delegated proxy credential from site A. The MyProxy java client running on site B calls the MyProxy get method as `get(username,password,lifetime)`. The following steps occur behind a `get` call.

   (a) The MyProxy server looks for the stored proxy credential using the username and verifies it using the password. If the username/password provided are valid, success is indicated to the java client at site B.

   (b) The site B then generates a public/private key pair similar to the one done in step 1 and sends a certification request to the MyProxy server.

   (c) The MyProxy server then signs the certification request using the private key of the stored proxy credential and returns a proxy certificate to site B.

4. The site B can then use proxy credential to access the GridFtp service running on site C. After accessing the GridFtp service, site B can send a `destroy` command to the MyProxy server. This will remove the credential from the server. This can be done as an added security measure so that the credential cannot be accessed by anyone else. The credential, username, password and lifetime are passed as arguments to the `destroy()` method.

MyProxy provides support for multiple authentication mechanisms. These include the basic pass-phrase method, Kerberos, PAM, SASL, and certificates. In the pass-phrase method the authentication is done by providing a pass-phrase entered while generating the proxy credential. PAM is an acronym for *pluggable authentication modules*. Using PAM, a Myproxy server can be configured to use external authentication such as using an LDAP server. SASL stands for *simple authentication and security layer*. Using SASL, MyProxy can be configured to provide support for standard authentication protocols like Kerberos.

4.3.3   Proxy Certificates

We have discussed the use of proxy certificates in Section 4.2.1.3 on GSI. To recall quickly, proxy credentials are used for single sign-on and dynamic delegation of privileges to other users or processes. In this section we discuss the detailed format of a proxy certificate and how it is generated.

4.3.3.1   Format of a Proxy Certificate

A proxy certificate follows the same format as the X.509 public key certificates. This ensures that the applications that make use of the X.509 certificates can also work with proxy certificates without any modification. Just like the X.509 certificate, a proxy certificate binds a unique name to a subject. A proxy certificate is identified by the end entity's certificate or another proxy certificate unlike the entity's certificate, which is verified using a certifying authority's certificate.

To achieve uniqueness, the subject name of a proxy certificate is prefixed by the subject name of the issuer [88]. For this discussion we term the subject name of the proxy certificate without the subject name of the owner as absolute subject name. The proxy certificate also contains a serial number just like the X.509 certificate. Both absolute identical name and serial number of the proxy certificate are generated using the hash of the public key to achieve uniqueness [89].

The public key of the proxy certificate is different from the original certificate because a new public/private key pair is generated for every proxy certificate. Other attributes of the key such as its size or algorithm used for generation may be different than the original public key.

The proxy certificate RFC [90] defines a field, `ProxyCertInfo`, which is an extension of the X.509 certificates. This extension indicates that this X.509 certificate is a proxy certificate. It also mentions the restrictions placed on this certificate by the issuer, if any. The `ProxyCertInfo` field contains two subfields `pCPathLenConstraint` and `proxyPolicy`. The value of `pCPathLenConstraint` field, if present, states the maximum depth of the path of proxy certificates that can be generated from this proxy certificate. If the field is absent, then the length of the path of the proxy certificates that can be generated from this proxy certificate is unlimited. The `proxyPolicy` field contains two fields. A `policy` field is an expression of policy specifying the use of this certificate for authorization. The other field is the `policyLanguage` field indicating the language used to express the policy. XACML and XrML can be used as policy language to express the delegation policies. The `policyLanguage` field can contain two values that are of particular significance and should be understood by all the applications working with proxy certificates. If the `policyLanguage` contains either of the values described below, the `policy` field must be absent. These values are:

1. id-ppl-inheritAll - This policy means that the issuer intends to delegate all of its privileges to the proxy certificate bearer.

2. id-ppl-independent - Through this policy type the issuer intends to provide a unique identifier to the proxy certificate holder. No issuer privi-

leges are granted through this.

Besides these policy types other policy types may be implemented, for example to delegate a subset of privileges to the proxy certificate owner.

4.3.3.2 Generation of Proxy Certificates

First we discuss the generation of proxy certificates for the purpose of single sign-on. The need for single sign-on arises because the private key of the user's X.509 end entity certificate is generally protected by a password. So it becomes cumbersome to access the private key each time a mutual authentication is required. Hence proxy certificates provide a convenient solution for single sign-on. In the first step, a new public/private key pair is generated for the proxy certificate and a certification request is created. The next step requires the access to the user's long-term private key associated with its X.509 end entity certificate. This generally requires entering a password because the long-term private key is kept secure. The private key is then used to sign the certification request generated in the first step creating the proxy certificate. The proxy certificate and the associated private key can be stored together on the local filesystem or stored on a MyProxy server as explained in the Section on managing credentials. Proxy certificates can also be created to delegate privileges from the issuer to another entity over a secure network connection, without exchange of private keys. This has already been explained in the section on managing credentials, where a MyProxy server delegates a proxy credential to a site to access the services on behalf of the owner of the X.509 end entity certificate.

References

[65] Laura Pearlman, Von Welch, Ian Foster, and Carl Kesselman. A community authorization service for group collaboration. In *Proceedings of the IEEE 3rd International Workshop on Policies for Distributed Systems and Networks*, pages 50–59. IEEE, 2002.

[66] R. Alfieri, R. Cecchini, V. Ciaschini, L. dell Agnello, A. Frohner, A. Gianoli, K. Lorentey, and F. Spataro. *VOMS: an authorization system for virtual organizations*, pages 33–40. Lecture Notes in Computer Science. Springer, March 2004. Available online at: `http://grid-auth.infn.it/docs/VOMS-Santiago.pdf` (accessed January 1st, 2009).

[67] A. Arenas. State of art survey on trust and security in grid computing systems. Technical report, Council for the Central Laboratory of the Research Councils, March 2006. Available online at: `http://epubs.cclrc.ac.uk/bitstream/909/RAL-TR-2006-008.pdf` (accessed January 1st, 2009).

[68] Audun Josang, Roslan Ismail, and Colin Boyd. A survey of trust and reputation systems for online service provision. *Decision Support Systems*, 43(2):618–644, 2007. Available online at: `http://sky.fit.qut.edu.au/~josang/papers/JIB2007-DSS.pdf` (accessed January 1st, 2009).

[69] Tyrone Grandison and Morris Sloman. A survey of trust in Internet applications. *IEEE Communications Surveys and Tutorials*, 3(4), September 2000. Available online at: `http://pubs.doc.ic.ac.uk/TrustSurvey/` (accessed January 1st, 2009).

[70] Ian Foster, Carl Kesselman, and Steven Tuecke. The anatomy of the grid: enabling scalable virtual organizations. In *Euro-Par 2001: Proceedings of the 7th International Euro-Par Conference, Manchester, UK, August 28-31, 2001*, volume 2150 of *Lecture Notes in Computer Science*, pages 1–4. Springer, August 2001.

[71] Network Working Group. Internet X.509 public key infrastructure certificate and CRL profile. Technical report, The Internet Society, January 1999. Available online at: `http://www.ietf.org/rfc/rfc2459.txt` (accessed January 1st, 2009).

[72] Roger M. Needham and Michael D. Schroeder. Using encryption for authentication in large networks of computers. *Communications of the ACM*, 21(12):993–999, December 1978.

[73] Kerberos. The Kerberos network authentication service v5. Technical report, Network Working Group, The Internet Society, September

1993. Available online at: `http://www.ietf.org/rfc/rfc1510.txt` (accessed January 1st, 2009).

[74] Roberto Alfieri, Roverto Cecchini, Vincenzo Ciaschini, Luca dell' Agnello, Akos Frohner, Karoly Lorentey, and Fabio Spataro. From gridmap-file to VOMS: managing authorization in a grid environment. *Future Generation Computer Systems*, 21(4):549–558, April 2005.

[75] John Hughes and Eve Maler. Security Authentication Markup Language (SAML) 2.0 technical overview. Technical report, OASIS, Security Services Technical Committe, February 2005. Available online at: `http://www.oasis-open.org/committees/download.php/11511/sstc-saml-tech-overview-2.0-draft-03.pdf` (accessed January 1st, 2009).

[76] Rick Randall and Booz Allen Hamilton. SAML X.509 authentication-based attribute sharing profile. Technical report, OASIS, Security Services Technical Committe, February 2005. Available online at: `http://www.oasis-open.org/committees/download.php/11323/sstc-saml-x509-authn-based-attribute-protocol-profile-2.0-draft-02.pdf` (accessed January 1st, 2009).

[77] Globus Alliance. Globus Toolkit v4: grid security infrastructure - a standard perspective. Web Published, September 2005. Available online at: `http://www.globus.org/toolkit/docs/4.0/security/GT4-GSI-Overview.pdf` (accessed January 1st, 2009).

[78] Bob Atkinson, Giovanni Della-Libera, Satoshi Hada, Maryann Hondo, Phillip Hallam-Baker, Chris Kaler, Johannes Klein, Brian LaMacchia, Paul Leach, John Manferdelli, Hiroshi Maruyama, Anthony Nadalin, Nataraj Nagaratnam, Hemma Prafullchandra, John Shewchuk, and Dan Simon. Web Services Security (WS-Security) v1.0. Technical report, IBM, Microsoft, Verisign, April 2002. Available online at: `http://www.verisign.com/wss/wss.pdf` (accessed January 1st, 2009).

[79] Takeshi Imamura, Blair Dillaway, and Ed Simon. XML encryption syntax and processing: W3C recommendation. Technical report, World Wide Web Consortium, December 2002. Available online at: `http://www.w3.org/TR/xmlenc-core/` (accessed January 1st, 2009).

[80] Hal Lockhart, Steve Andersen, Jeff Bohren, Yakov Sverdlov, Maryann Hondo, Hiroshi Maruyama, Anthony Nadalin, Nataraj Nagaratnam, Toufic Boubez, K. Scott Morrison, Chris Kaler, Arun Nanda, Don Schmidt, Doug Walters, Hervey Wilson, Lloyd Burch, Doug Earl, Siddharth Baja, and Hemma Prafullchandra. Web Services Federation language (WS-Federation) v1.1. Technical report, BEA Systems et al., December 2006. Available online at: `http://specs.xmlsoap.org/ws/2006/12/federation/ws-federation.pdf` (accessed January 1st, 2009).

[81] Siddharth Bajaj, Don Box, Dave Chappell, Francisco Curbera, Glen Daniels, Phillip Hallam-Baker, Maryann Hondo, Chris Kaler, Dave Langworthy, Anthony Nadalin, Nataraj Nagaratnam, Hemma Prafullchandra, Claus von Riegen, Daniel Roth, Jeffrey Schlimmer, Chris Sharp, John Shewchuk, Asir Vedamuthu, Ümit Yalcýnalp, and David Orchard. Web Services Policy framework (WS-Policy) v1.2. Technical report, BEA Systems et al, March 2006. Available online at: `http://specs.xmlsoap.org/ws/2004/09/policy/ws-policy.pdf` (accessed January 1st, 2009).

[82] Steve Anderson, Jeff Bohren, Toufic Boubez, Marc Chanliau, Giovanni Della-Libera, Brendan Dixon, Praerit Garg, Martin Gudgin, Phillip Hallam-Baker, Maryann Hondo, Chris Kaler, Hal Lockhart, Robin Martherus, Hiroshi Maruyama, Anthony Nadalin, Nataraj Nagaratnam, Andrew Nash, Rob Philpott, Darren Platt, Hemma Prafullchandra, Maneesh Sahu, John Shewchuk, Dan Simon, Davanum Srinivas, Elliot Waingold, David Waite, Doug Walter, and Riaz Zolfonoon. Web Services Trust language (WS-Trust). Technical report, BEA Systems et al, February 2005. Available online at: `http://specs.xmlsoap.org/ws/2005/02/trust/WS-Trust.pdf` (accessed January 1st, 2009).

[83] Ian Foster, Carl Kesselman, Jeffery M. Nick, and Steven Tuecke. The physiology of the grid: an open grid services architecture for distributed systems integration. Web Published, 2002. Available online at: `http://www.globus.org/alliance/publications/papers/ogsa.pdf` (accessed January 1st, 2009).

[84] Security architecture for open grid services. Technical report, OGSA, July 2002. Available online at: `http://www.globus.org/toolkit/security/ogsa/draft-ggf-ogsa-sec-arch-01.pdf` (accessed January 1st, 2009).

[85] OGSA. Global grid forum specification roadmap towards a secure OGSA. Technical report, OGSA Security Workgroup, July 2002. Available online at: `http://www.globus.org/toolkit/security/ogsa/draft-ggf-ogsa-sec-roadmap-01.pdf` (accessed January 1st, 2009).

[86] Globus Alliance. Globus Toolkit 4.0.5. Web Published, 2007. Available online at: `http://www.globus.org/toolkit/downloads/` (accessed January 1st, 2009).

[87] Jim Basney. MyProxy protocol, November 2005. Available online at: `http://www.ggf.org/documents/GFD.54.pdf` (accessed January 1st, 2009).

[88] Burton S. Kaliski. PKCS #10: certification request syntax. Technical report, Network Working Group, The Internet Society, March

1998. Available online at: `http://www.ietf.org/rfc/rfc2314.txt` (accessed January 1st, 2009).

[89] Von Welch, Ian Foster, Carl Kesselman, Olle Mulmo, Laura Pearlman, Steven Tuecke, Jarek Gawor, Sam Meder, and Frank Siebenlist. X.509 proxy certificates for dynamic delegation. Web Published, 2007. Available online at: `http://www.globus.org/alliance/publications/papers/pki04-welch-proxy-cert-final.pdf` (accessed January 1st, 2009).

[90] Steven Tuecke, Von Welch, Douglas Engert, Laura Pearlman, and Mary R. Thompson. Internet X.509 Public Key Infrastructure (PKI) proxy certificate profile. Technical report, Network Working Group, The Internet Society, June 2004. Available online at: `http://www.ietf.org/rfc/rfc3820.txt` (accessed January 1st, 2009).

Chapter 5

Grid Middleware

5.1 Overview of Grid Middleware

Grid systems include an aggregation of computational resources, storage resources, network resources and scientific instruments. These resources construct the so-called grid fabric of the grid [106]. Computational resources may be supercomputers, servers, dedicated computers, PCs or clusters existing in several geographically distributed organizations. Scientific instruments such as telescopes and sensor networks provide real-time data that can be transmitted directly to computational sites or that can be stored in a database. These instruments are most of the time distributed, heterogeneous and hybrid. To effectively use these resources, we need some middleware for accessing grid resources, facilitating usage of these resources, and protecting the interest of resource providers.

Grid middleware provides users with seamless computing ability and uniform access to resources in a heterogeneous grid environment. The development of grid middleware should be shared, reusable and extensible [107] to minimize the time needed for developing and deploying it again. Grid middleware is a collection of APIs, protocols, and software that allow creation and use of a grid system. However, as a grid is commonly used within one or several communities, and developing just the APIs as a part of the middleware is possible, it does not provide the abstraction level needed for practical grid projects. Because of the complexity of the grid architecture and the diversity of its function, the interaction between middleware and other software is also required. For these practical requirements, a grid middleware has a layered architecture.

In the architecture of a grid, the bottom layer is the grid fabric layer composed of the grid resources, while the top layer is the user's grid application layer. The grid middleware layer is located between these two layers. A fourth layer architecture of grids is also described in [105]. It divides the grid middleware layer into two distinct layers. According to this approach, the whole architecture includes: grid fabric, core grid middleware, user-level grid middleware, grid applications and portals. Figure 5.1 illustrates the grid middleware in the grid architecture.

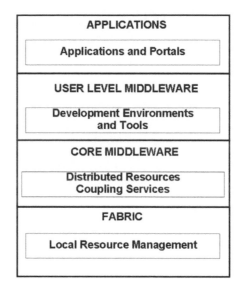

FIGURE 5.1: Grid middleware in the grid architecture.

In this architecture, the middleware layer is divided into two levels: *core grid middleware* and *user level grid middleware*. Core grid middleware offers the elementary services such as remote process management, co-allocation of resources, storage access, information registration and discovery, security, and aspects of Quality of Service (QoS) such as resource reservation and trading. These services abstract the complexity and heterogeneity of the fabric layer by providing a consistent method for accessing distributed resources. User level grid middleware is based on the interfaces provided by the low-level middleware to construct higher level abstractions and services. These include application development environments, programming tools and resource brokers for managing resources and scheduling application tasks for the execution on global resources [105]. In the following section, the services in these two levels will be described in detail.

5.2   Services in Grid Middleware

Grid middleware components provide the necessary functions for the users to utilize the grid. These functions include authentication, authorization, resource matchmaking, data transfer and monitoring. This is the first level of the grid middleware functions. More sophisticated services such as resource management, resource broker, job management, information management, etc., can be developed on top of these elementary functions.

5.2.1   Elementary Services

The elementary services of a grid middleware include *job execution, security, information service* and *file transfer*.

Job execution supports the submitted jobs running on a remote resource. The security of the resource is guaranteed with appropriate security services providing the authorized utilization of the grid resource. The security mechanism is concerned with the user's authentication, authorization and secure access to the grid resources. The support for asynchronous job execution is necessary. During the execution, the job submitter should be allowed to inquire about the execution status. This implies that the submitted jobs should have a unique identifier within the grid resource in order to be correctly distinguished and scheduled.

Grid security provides the authentication and authorization mechanism for users to access the grid resources in a secure way. These security mechanisms are based on the network security. The traditional network security protocols and methods are adopted by grid technologies, such as encryption, digital certificate and other elementary security mechanisms that are usually included in grid security services. Public Key Infrastructure (PKI) and Kerberos are two frequently-used authentication mechanisms. In addition to the basic network security mechanism, grid systems require the *single sign-on* feature to facilitate the job execution that requires access to the grid resources across several sites. Users have to enter their passwords only once to access to all the required resources. For this purpose a proxy credential is used.

Information service. A grid is a dynamic environment within which the status of the resources and the location of the service change dynamically. An information service provides the ability to query the information about resources and services, for instance, the utilization of CPU, the amount of free memory of a computer, the availability of services, etc. This information is recorded and analyzed enabling higher level functions like job execution. An information service may refuse an unreasonable request from the user. Nowadays, traditional semantic web is being incorporated into grid computing. It introduces a new term semantic grid. The metadata of data, service or object, combined with information services, offers users a more effective means to query the resources.

File transfer offers an ability to safely and efficiently transfer the data needed for a task from its original location to the node where the computation is done. The user would like a simple command to accomplish this task. For example, "copy file A from server X to file B on server Y as fast as possible" [107]. Some features such as striping and parallel streams are included in file transfer, when the data being transferred is large.

5.2.2   Advanced Services

On top of elementary services provided by grid middleware, a variety of more sophisticated services can be developed. *File management, information management, job management* are three main examples of such services.

File management gives users the ability to control the file transfer and monitor its status. Grid middleware should offer these functions and hide the underlying complexity from the user. The replica service is a method of increasing the data redundancy in order to reduce the cost of file transfer. A set of replica services are provided by the middleware. These include location of data replica, replica management, etc.

Information management gives information about storage and computational resources in the grid as well as the information published by other grid components. It also includes the status of executing tasks as they are collected, stored and analyzed. Information about the resources and their access pattern provides the user with a global view of the grid resource consumption. Further, information service is usually coupled with a *monitoring service*. Information and monitoring services help in discovering the resources and finding their properties and thus provide an important service in a grid system.

Job management is another important service that deals with the coordination of the services offered by security, file transfer and information service components. The job submitted by the user may consist of many subtasks and require the use of large-scale data. It may also be computationally intensive in terms of number of hours required by it. For successful execution of such jobs, a fault tolerance mechanism is also necessary.

The previous discussion is an abstract of the middleware services for clearly distinguishing their functionality. From a real point of view, these services are closely associated with each other.

5.3   Grid Middleware

Many research projects have been undertaken for the development of a grid middleware toolkit. A lot of significant software has been designed and realized. They address some of the complex issues related to the grid. Some of them have been designed to provide all the functions of the grid ranging from the basic service to higher-level ones. Others just focus on narrow aspects. In this section, we introduce the commonly used grid middleware, and classify them according to their functionalities.

5.3.1   Basic Functional Grid Middleware

We describe the middleware keeping in mind the fundamental functions it provides. Globus Toolkit (GT) is an important middleware and the *de facto* standard for grid implementation. It will be the focus of our discussion later in the chapter. In this section, we present two vertically integrated middleware products [105] namely UNICORE and Legion. They provide the most important components needed in grid middleware. They provide functions ranging from the resource access concerned with the grid fabric layer to the user-level components dealing with the grid application.

5.3.1.1   UNICORE Middleware

UNICORE [91] is the outcome of the project UNICORE (UNiform Interface to COmputing REsources), which has created a technology that provides seamless, secure, and intuitive access to distributed grid resources. Its objectives include the development of a prototype supporting multisite job execution and an application-specific interface. It also allows efficient data transfer, resource modeling and meta-computing at the application level.

Unicore has been implemented on top of web technologies. It uses Java and HTML, which are the defacto standard for displaying web pages language. DNS and LDAP offer uniform access to distributed directories of files, names and other resources. A client/server protocol, https, provides secure access and transfer of information. Certification provides the basis for reliable and uniform authentication mechanism for users and other subjects in the grid.

In Figure 5.2, the client side contains an element, *Job Preparation Agent (JPA)*, a Java applet running within the user's web browser from where a job *Abstract Job Object (AJO)* has been created. The user can define an AJO and specify a target machine where the user has a registered account. After an AJO is constructed, it will be passed on to the *Network Job Supervisor (NJS)* with the user's certificate through a *gateway*. Both NJS and gateway run on the target site. The *Job Monitor Controller (JMC)* running on the client side is used to obtain the job status and the final results.

The gateway provides a single point of entry to the UNICORE *Usite* system. It consists of a `https-capable-server` and a Java servlet. It provides secure authentication and authorization to the user. After authenticating the user's UNICORE certificate, the server passes the AJO to the servlet. The servlet extracts the site-independent identity (*Ulogin*) and searches for the information associated with the user in the *user database* to check the user's privileges to access the required resources. Then the servlet maps the *Ulogin* to a local Unix login (*Xlogin*). After this the AJO is sent to the NJS.

UNICORE *Vsite* consists of *Network Job Supervisor (NJS)* and *Target System Interface (TSI)*. NJS receives AJO and begins to process it. Meanwhile, NJS sends an acknowledgment to JPA. After this the user can quit the system or submit another job. During the execution, a *Uspace* is created to store all the files, including those needed for execution and the files created by AbstractTask. NJS translates the AbstractTask into a set of commands that can be executed by *Target System Interface (TSI)*. During this translation procedure, translating tables in the *incarnation database* are used to map the abstract representations to concrete commands. TSI accepts incarnated job components from the NJS, and passes them to the local batch systems for execution. The child AJOs embedded in the AJO are relayed to the specified target site. Similar processing is done on them. When the work defined by the AJO is finished, an acknowledgement is returned to the user and the *Uspace* is deleted.

Compared to other grid platforms, an advantage of UNICORE is its friendly and powerful client software. It supports all the versions of Java and can be installed under any version of Unix (Linux included). For data management, UNICORE constructs a temporary data space for every job for importing and exporting the data. It permits remote file browsing and automatically executes the data transfer when the job terminates. UNICORE offers a unified interface to computational resources. Gateway, NJS and TSI form the UNICORE server, called Usite. NJS and TSI form the Vsite, which hides the disparities among various platforms thereby offering a uniform interface. UNICORE mainly addresses batch jobs and does not support jobs that require user interaction. If there exists interaction among the user's program, UNICORE cannot correctly perform the task.

UNICORE technology serves as a solid base in many European and international research projects, such as EuroGrid and their applications namely, BioGrid, MeteoGrid, CAEGrid, Grid Interoperability Project (GRIP), Open-MolGrid, and Japanese NAREGI (National Research Grid Initiative). UNICORE is used as their low-level middleware. In addition, the integration of

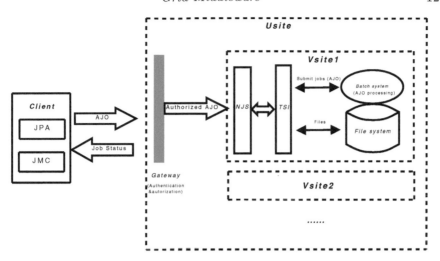

FIGURE 5.2: Architecture of UNICORE middleware.

UNICORE into grid service has been attempted in Unicore/GridService [108]. In this, UNICORE acts as a scheduling component, called the *Unicore server*. Unicore Server receives AJOs from the *Unicore client*, the grid service platform, and other Unicore servers. These Unicore servers are accessed using the *Unicore Protocols Layer (UPL)*.

5.3.1.2 Legion Middleware

Legion was created by the University of Virginia, USA. It envisions that every user and user group should have a visual computer with high computing performance via parallelism, huge storage space and easy access to distributed data and resources. Legion is also an open source project so that it can easily adapt to the rapid change of software and hardware. It offers a uniform namespace. It supports applications written in different programming languages. Legion provides a single, coherent virtual machine that addresses issues such as scalability, programming ease, fault tolerance, security, and site autonomy. Legion is a conceptual base for a metasystem [105].

Figure 5.3 shows the architecture of the Legion middleware. Legion runs on the top of the user's operating system and links the user's hosts and other required resources. Legion acts as a negotiator for scheduling and security policies on behalf of the user. Context-space is a user-controlled naming system, which allows users to easily create and use the objects in remote systems. The Legion object allows the execution of jobs to be carried out in heterogeneous platforms. Users can run applications written in multiple languages, since Legion supports interoperability between objects written in multiple lan-

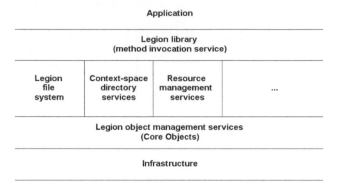

FIGURE 5.3: Architecture of Legion middleware.

guages [98].

While many research projects have addressed problems related to scheduling computational resources on parallel computers or local networks, few have addressed the distinctive problems that arise when resources are distributed across many sites. Legion is an exception in this regard. Indeed, Legion supports the access to distributed and heterogeneous computing resources, and it can keep track of the availability of the binary code and of the current version.

Legion implements several core objects that offer the main services provided by Legion middleware. The class object is used to define and manage every Legion object. Class object is responsible for creating a new instance and scheduling it for execution. It also activates and deactivates objects and provides location of an object to other objects that want to communicate with it. The other core objects include host, vault, context, binding agent, implementation, implementation cache. Host and vault are two kinds of resources. Host represents the computational resource, while vault represents the storage resource that maintains the object's state. Context and binding agents serve for the Legion naming system. Context maps context names to the Legion Object Identifiers (LOIDs). Binding agents bind object addresses, Legion Object Addresses (LOAs), to LOIDs on behalf of other Legion objects. Implementation maintains an executable file for requests to activate or create an object. Implementation cache stores the implementation so that it can be reused by multiple host objects to avoid storage and communication costs. Figure 5.4 illustrates some Legion objects in the Legion middleware.

As Legion is an object-oriented middleware, every resource is an object-file. Data resources are also represented by objects. *BasicFileObject* represents a

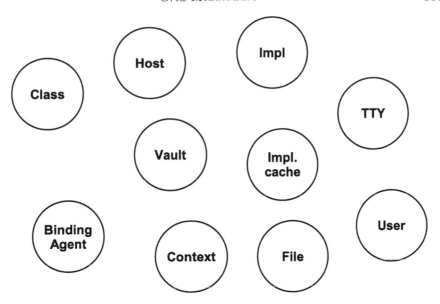

FIGURE 5.4: Legion objects in Legion middleware.

common file in conventional file systems, and *ContextObject* represents the directory. This representation has its own advantages. As these file objects are registered in Legion context, they exist in single-addressed space. Further, this representation of the data file is independent of the system allowing inter-operability between heterogeneous systems. For accessing data objects, Legion offers three principal means: *command-line interface, user API* and *buffering interface*. Modifications of the local copy of the Legion objects and the changes are copied to the object itself. These three data access means provide a transparent I/O mechanism for accessing files stored on distributed nodes.

Because of the rapid growth in IT infrastructure such as network bandwidth, CPU processing speed, larger memory and disk space, more and more complex applications are being deployed using Legion. A new business-oriented software named Avaki has been founded on the basis of Legion to serve such needs. In Avaki, the original architecture of Legion is retained. Error-checking ability is added for reinforcing the robustness of the infrastructure. The architecture of Avaki is similar to the architecture of OGSA proposed by GGF. All objects (called grid resources in OGSA) have a name, an interface, a way to discover the interface, metadata and state, and are created by factories (analogous to Avaki class objects) [109]. Some significant proposals have been made by Avaki. For example the proposal of Secure Grid Naming Protocol (SGNP) to the GGF as an open standard for naming in grids. This demonstrates the contribution of Avaki in the standardization of grid technology.

5.3.2 High-Throughput Computing Middleware

High-performance computing pays attention to the number of jobs completed in a short period of time. Engineers and scientists frequently face problems that need weeks or months of computation to be solve. They are not directly concerned with operations at the scale of one second but operations at the scale of a month or a year. These problems involve a larger scale and users are mainly interested in how many jobs they can complete over a long period of time instead of how fast an individual job can be completed. This type of computation is called *high-throughput computing (HTC)*.

A high-throughput computing system is a system optimized to increase the utilization of resources. It should have the following important qualities: high performance, fault tolerance and robustness. In the following paragraphs, we describe some middleware well suited for high-throughput computing.

5.3.2.1 Condor Middleware

Condor is a software system that creates a HTC environment. It is a local scheduling system for computationally-intensive jobs. It effectively utilizes the computing power of idle workstations connected by communication network.

Condor provides a job queueing mechanism, scheduling policy, priority scheme, resource monitoring, and resource management. Once the user submits a job, Condor puts this job into a queue, and chooses when and where to execute the job according to the scheduling policies and priorities as well as the resource requirement given by the user. When the job is finished, Condor informs the user about the completion.

Compared to other batch systems, Condor provides two features: high-throughput computing and opportunistic computing [104]. A high-throughput system requires the support of fault tolerance. The checkpointing and migrating mechanisms of Condor fulfill this necessity. Checkpointing means taking the periodical snapshot of the job during its execution. This snapshot allows Condor to resume a failed task by rescheduling it to another machine thus providing fault tolerance. Opportunistic computing means that Condor can utilize the otherwise wasted CPU capacity. If a dedicated node does not have a job to execute (is idle), Condor can utilize this node in an opportunistic manner.

Condor is composed of a set of components, which accomplish the kernel functions of Condor, as represented in the Figure 5.5. The *agent* receives the job submitted by users, each job has its own agent. The *resources* and agents advertise themselves to the *matchmaker*, which finds the jobs and the potentially compatible resources. Condor offers two *problem solvers* for solving complex problems: the *masterworker (MW)* and the *directed acyclic graph manager (DAGMan)*. Problem solver is a bridge between a user and the agent. MW addresses the problem with unknown size and unreliable workforce. DAGMan is a service for executing multiple jobs with dependencies in a declarative form.

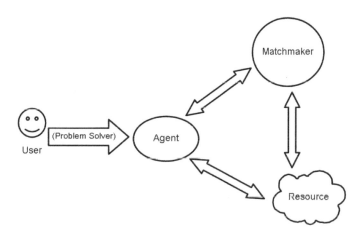

FIGURE 5.5: Main components of Condor middleware.

The *ClassAds* mechanism in Condor provides an extremely flexible, expressive framework for matchmaking resource requests with resource offers. This is a language for describing the user's jobs and the resources (i.e. the machines). The ClassAds file is a list of expressions of attribute and value (similar to C expressions) having the following format: `Owner=''Mike''`. The job's ClassAds includes requirements about the machines with which it should be matched. Examples of such requirements are free disk size, minimum memory size, etc. The machine's ClassAds describes the resource attributes and the features it can provide to a particular job. The ClassAds mechanism is finally used to match the resource requirements and the resource offers.

In Condor, a user indicates the job execution environment by specifying the attribute universe in the job description file. Condor provides several predefined values of universe for different user jobs: standard, vanilla, PVM, MPI, grid, Java, scheduler, local, parallel. If the user doesn't specify a value for universe, standard is used as the default universe. Standard universe supports migration and reliability, but has some restrictions on the programs that can be run. The Java universe allows users to run jobs written for the Java Virtual Machine (JVM). The characteristic of other universes can be found in reference [103]. Universe is used to match shadow and sandbox, two major processes of Condor running at the agent and the resources, respectively [109]. Shadow is responsible for providing all the details necessary to execute a job in the agent. Sandbox is responsible for creating a safe execution environment for the job and protecting the resource from any mischief in the resource. Universe is a linkage of these two process. So developing a new universe requires the deployment of new software modules at both the ends.

The *Condor pool* is a set formed by agents, resources, and matchmakers. Utilizing the Condor pool, remote resources can be collected; furthermore, the scheduling spanning sites can be realized. From the user's point of view, the Condor is somewhere to submit the job. The users send the job to the Condor pool, and wait until the jobs are staged on the pool. They can then disconnect from the pool. When they reconnect they can get the job results if the job has finished execution.

Condor-G is the grid version of Condor. It is the combination of Condor and Globus. The architecture of Condor-G includes Globus Toolkit components, such as GRAM, GSI and MDS. A GridManager is locally created in each resource node for controlling the local tasks, searching the available resources and scheduling the tasks on the possible resources. The GSI and MDS take charge of the security and information services respectively.

As shown in Figure 5.6, Condor-G is used as the submission and job management service for one or more sites. The Condor high-throughput computing system can be used as the fabric management service (a grid generator) for one or more sites. The Globus Toolkit services can be used as the bridge between them.

5.3.2.2 Nimrod-G Middleware

Nimrod-G is a high-throughput computing system implemented on Globus Toolkit. The idea of computational economy is taken into account in the scheduling policy of Nimrod-G. Users can specify the deadline of the task execution and the budget for their tasks. Their applications can be scheduled according to their budget specification. Nimrod-G tries to execute the applications within a given deadline and with a given cost. The deadline represents a time by which the user requires the result. Such requirements are largely limited to parameter sweep (or task farming) applications.

Nimrod-G is a tool for automated modeling and execution of parameter

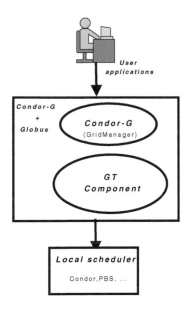

FIGURE 5.6: Condor and Condor-G Globus middleware.

sweep applications over global computational grids. The parameter sweep application is described as an application model as part of a high-throughput computing (HTC) application [110]. This model is a combination of task and parallel data model. The applications formulated using this model contain a large number of independent jobs operating on different data sets. The programming and the execution model of such applications are similar to the Single Program Multiple Data (SPMD) model, which involves processing multiple independent jobs (each with the same task specification, but a different dataset) on distributed machines. The number of jobs is usually much larger than the number of machines. PSAs are used for molecular modeling of drug design, protein folding, searching for extra-terrestrial intelligence, human-genome sequence analysis, brain activity analysis, high-energy physics events analysis, ad hoc network simulation, crash simulation, financial modeling, and Mcell simulations.

Nimrod-G acts as a resource broker in the grid architecture. It utilizes GRACE [111] services for identifying and negotiating low cost access to computational resources. It consists of four components: a task farming engine, a scheduler, a dispatcher and resource agents. The task farming engine is a job control agent that manages and controls an experiment. It is responsible for the entire experiment, from its parametrization to the actual job creation and the job status maintenance. The scheduler takes charge of resource discovery, resource trading, resource selection and job assignment. It is able to allocate grid resources and services depending on their capability, cost, and availability. The dispatcher uses Globus for deploying agents on remote resources to manage the execution of assigned jobs. Agents are deployed on grid resources dynamically at runtime according to the instructions of the scheduler. It is responsible for the environment building for a job in a given resource, data transferring, startup of task and the return of results. Nimrod-G integrates the components of Globus Toolkit. It uses Metacomputing Directory Service (MDS) services for resource discovery. Globus Resource Allocation Manager (GRAM) is used for starting the tasks on a remote computational resource and managing the execution of tasks. Grid Security Infrastructure (GSI) provides the mechanism of single sign-on on multiple computational resources for avoiding re-authentication when the computation concerns several resource sites. The Global Access to Secondary Storage (GASS) service allows users to access various storage systems with a uniform access mechanism. Scientists and engineers are able to model whole parametric experiments and transparently stage the data at remote sites.

The scheduling algorithm of Nimrod-G takes into account the users' cost and deadline requirements. The Nimrod-G adaptive scheduling algorithm is designed to minimize the time and/or the cost of computations for user-defined constraints. For example, the user can define the deadline by which the results are needed and the Nimrod-G broker will try to find the cheapest resource available in order to meet the user-defined deadline and keep the cost at the minimum. If the scheduler detects that the tasks cannot be completed before the deadline, then the scheduler abandons the current resource, tries to select the next cheapest resource and tries until the completion of task farm application meets the deadline. Nimrod-G provides four scheduling algorithms for this scenario:

- Cost optimization uses the cheapest resources to ensure that the deadline can be met and that computational cost is minimized.

- Time optimization uses all the affordable resources to process jobs in parallel as early as possible.

- Cost-time optimization is similar to cost optimization, but if there are multiple resources with the same cost, it applies the time optimization strategy while scheduling jobs on them.

- The conservative time strategy is similar to time optimization, but it guarantees that each unprocessed job has a minimum budget per job.

Unlike Condor, Nimrod-G adopts another approach to attain the objective of high throughput. In the scheduling model of Nimrod-G, the user can specify the budget and finish time (deadline) in the application requirements. The user can even negotiate access to computational resources at low cost during the scheduling. This economy driven approach enables us to build a scalable computational grid that follows a user-centric approach in scheduling. However, currently the applications of Nimrod-G are merely the parameter-sweep applications. For other scientific and numerical applications, novel scheduling algorithms based on economy should be researched.

5.3.3 GridRPC-Based Grid Middleware

GridRPC is a programming model based on client-server remote procedure call (RPC). It is a standardized, portable and simple programming interface for RPC over the grid. Ninf and NetSolve are two notable grid middleware based on the GridRPC model.

5.3.3.1 NInf/NInf-G Middleware

Network-based information library for high-performance computing (*NInf*) [96] is a client/server distributed programming platform, which is based on the remote procedure call (RPC) model. It aims to provide a platform for global scientific computing with computational resources distributed in a world-wide global network.

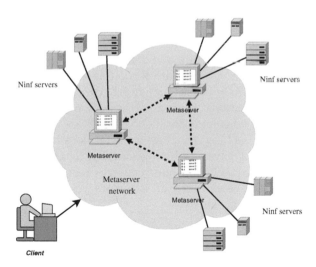

FIGURE 5.7: Architecture of Ninf: client, metaserver and server.

Figure 5.7 shows the Ninf components from its architectural point of view. The main components of Ninf include client, metaserver and server. The requests coming from the client are sent to the metaserver, which acts as an agent between the clients and servers. When metaserver receives a client request, it looks for the appropriate server to execute the required library. In Ninf, the program in the remote server exists as a library. These remote libraries form the computational resources in Ninf. The request from the client is transformed into a call for the execution of the remote library.

Executable libraries are the computational resources in Ninf. Some Ninf libraries such as LAPACK have been predefined and implemented. Users can also create their own library according to their needs. The creation of the library begins with the description of the interface of the library function in *Ninf IDL*. Then Ninf IDL and the library are compiled and linked to the Ninf RPC. Finally, the library should be registered to the Ninf server. Thus, any user can use a library registered to the Ninf RPC. The invocation of a remote library is simple. All we have to do is to add the function `Ninf\_call()` with appropriate parameters. The first parameter is the library name, followed by the input and output parameters of this library.

Ninf Metaserver is an agent providing functions similar to the grid information service. It gathers network information about the Ninf servers and helps the client to choose an appropriate Ninf server. When choosing the Ninf server, the metaserver considers load balance among different Ninf servers for optimal selection and better utilization of available servers.

Ninf has the following advantages. Thanks to the language-independent design of Ninf, the client program and remote library can be written in any language. Ninf uses the Sun XDR data format as a default protocol to resolve the challenge of data format consistency in heterogeneous environments. Ninf supports parallel applications. During the execution of a parallel application, the data dependencies among the parallel Ninf calls are automatically detected and scheduled by the metaserver. However Ninf does have some limitations. Currently, Ninf doesn't have a robust security mechanism. Fault tolerance mechanisms (involving checkpoint and recovery) reduce the execution speed. A client's code cannot be executed on a remote host directly. It affects the efficiency of execution to some extent.

Ninf-G is the implementation of Ninf on top of Globus Toolkit. It is developed using Globus C and Java API. The GT components, GSI (Grid Security Infrastructure), GRAM (Globus Resource Allocation Manager), MDS (Monitoring and Discovering Service), GASS (Global Access to Secondary Storage) and Globus-IO are employed in Ninf-G. Some functionalities in old Ninf are taken over by these GT components. For example, the interface information (LDIF file[1]) retrieval is carried out using MDS (GRIS). The invocation of

---

[1] LDIF file is used for saving the interface information and registering in MDS.

remote library executable is performed by GRAM. With the combination of GT's components, Ninf-G offers a secure and inter-operable GridRPC framework. The performance of Ninf-G is affected due to the MDS lookups and GRAM invocation overheads.

5.3.3.2 NetSolve Middleware

NetSolve [93] is a distributed computing system developed by the University of Tennessee, USA. NetSolve is similar to Ninf. In Netsolve, there are three components: a client, an agent and a server. The client sends a request for a resource to the agent. The agent returns a list of available servers; the client tries to connect to one of these servers and sends data to the server. The server executes the client's task and then returns the results. Applications for NetSolve can be written in C and Fortran. It supports Matlab and Mathematica applications also. GridSolve, like NetSolve, is also a broker middleware. However, GridSolve can also be used as a proxy for the other grid services, such as Condor. Users don't need to know about the underlying services. Figure 5.8 illustrates the architecture of the NetSolve system.

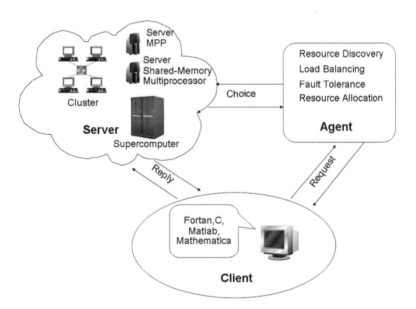

FIGURE 5.8: Architecture of NetSolve system.

Client Interface

The NetSolve client interface is implemented as a set of library routines called from a user's program. Users can invoke these library routines in their C or Fortran program. In C, the function to be called is `netsl()`, while in Fortran it is `FNETSL()`. As an example, the invocation of routines issued from the LAPACK[2] library is given in the NetSolve users' guide [94] as follow:

```
netsl( dgesv () ,n,1,a,lda,ipiv,b,ldb,&info);
```

where `dgesv()` is the routine to be called, followed by the list of arguments of this routine. The same routine invocation in Fortran is written as follow:

```
CALL DGESV (N,1,A,LDA,IPIV,B,LDB,INFO)
```

In addition, for users who are not familiar with compiled programming languages, Netsolve provides interactive interfaces such as a Matlab interface and a Mathematica interface.

NetSolve Agent

The NetSolve agents act as a resource broker. They are C programs running as stand-alone daemons. Only one agent can run on a machine at a given time. The agents periodically collect information about the servers and their status. Specifically, server information includes all the software or problems the server contains, the architecture of the server, work load of server, etc. The agents use a variety of mechanisms for retrieving this information; for example, standard Unix calls, such as `uptime`, are used to get the server load; Globus heart beat monitor is used to detect failures; network weather service is used to record the network status information. This information is stored in a database maintained by the agent. According to this information, the agent allocates server resources for client requests. The resource allocation decision takes into account factors such as execution speed and load balance. When the agent receives the client request, it identifies the server most appropriate for the request and returns its network address to the client library.

NetSolve Servers

The NetSolve server is the computational backbone of the system. It is a daemon process awaiting client requests. The server process may be running on a supercomputer, a massively parallel processor (MPP), a shared-memory multiprocessor or a cluster of workstations. The problems to be solved by the server are described or wrapped by the NetSolve Problem Description File (PDF). A key component of the NetSolve server is a source code generator

---

[2]LAPACK (`http://www.netlib.org/lapack/`). LAPACK is a library providing routines for solving linear systems of equations. LAPACK is one of the scientific packages integrated within NetSolve.

that parses the PDF and allows NetSolve to add new modules and function-alities. NetSolve servers are divided into two categories: the software servers (database or repository of software) and the hardware servers (containers for running the software). Users can register their computer as a software server or hardware server. Once they are registered, the computer is recorded into the hardware server list or the software list in the agent. When a client makes a request, the agent chooses the software server that has the relevant software and allocates the task to the appropriate hardware server according to the server's current workload.

Currently, the agents and servers run on all variety of Unix machines, and clients may run on Unix or Windows.

Security

Unlike Ninf, NetSolve has a security mechanism. It utilizes Kerberos for authentication. NetSolve includes three components, a client, an agent and a server. However, the client makes authentication requests only to the server [93]. A Kerberized client can access both a non-Kerberized server and a Kerberized server. When a non-Kerberized server receives a request, it returns a status code indicating that it can accept the request. The Kerberized server asks the sender to send its Kerberos credential for an authentication in response to the request.

Problem Description File

The Problem Description File (PDF) is a kind of Interface Definition Language (IDL). For a problem to be solved, there must exist a PDF corresponding to the problem. The PDF contains the information that allows NetSolve to create new modules and integrate new software, which specify the parameters, data types, and *calling sequence*[3]. When a new software is added, the server's problem list is appended. For facilitating the revising of the PDF, a user friendly GUI is provided, which can check various errors in user input and mostly consists of mouse clicks and choices in menus.

NetSolve is GridRPC-based middleware. In NetSolve2.0, an implementation of the GridRPC API is included. As Ninf is also supported by GridRPC, it is theoretically possible to unify NetSolve and Ninf by a common API. But currently this idea is still only in the imagination. *Task farming* is a mechanism to manage a large number of requests for a single NetSolve problem. This mechanism is very useful for the kind of algorithms that contain multiple independent tasks, for example, Monte Carlo simulations and genome sequences. Consequently, users should know how to schedule their tasks. But this needs the user to know the system details, which is not easy for ordinary

---

[3]The input and output parameters of a NetSolve problem are both objects. So, for non-object-oriented languages such as C and Fortran, which can not manipulate objects directly, it is necessary to use a *calling sequence* that describes the objects' features individually.

users. Further work is being done in GridSolve (the current developed version of NetSolve), for example, eliminating the limitation that users cannot withdraw results until all the tasks are completed.

5.3.4   Peer-to-Peer Grid Middleware

Peer-to-peer (P2P) is a special grid system. It couples heterogeneous resources such as computers using a variety of operating systems and networking components. One of the biggest challenges with P2P systems is enabling devices to find one another in a computing paradigm that lacks a central point of control. A P2P system needs the ability to find resources and services from a potentially huge number of globally located, decentralized peers. In addition, as the PCs or other individual resources connect to the Internet voluntarily, the resource pool formed by these PCs is not constant. So, fault tolerance is also a typical challenge in P2P systems.

5.3.4.1   XtremWeb Middleware

XtremWeb [97] is a P2P platform for grid computing experiments developed in Java. The goal of XtremWeb is to transparently exploit networked resources on a large geographic scale through the Internet. XtremWeb belongs to the so-called Cycle Stealing Environment family. That is, XtremWeb utilizes the CPU idle time of computational resources (PCs, workstations, PDA, servers) connected by the Internet.

As demonstrated in Figure 5.9, there are three main components in the architecture of XtremWeb: *Server*, *Worker* and *Client*. The tasks submitted by clients are registered on the server and are scheduled on the workers. During the execution, the workers contact the server to get the tasks. The server sends a parameter set and the concerning application in the response. When the workers finish their tasks, they contact the *collector* to return their results. This architecture adds a middle tier between client and work nodes. There is no direct P2P task submission/result transfer between clients and workers. The third tier, server, decouples clients from workers and coordinates the task execution on workers.

XtremWeb makes use of Sandboxing to securely perform the task. Sandboxing puts the running executable in a secure and closed context. So even if the execution is hostile to the running context, the resource provider's system remains safe. The most commonly known sandbox mechanism is the Java Virtual Machine, which interprets a byte-code. All interactions between the program and the system pass through the JVM, which filters them. XtremWeb is implemented using Java and so it is naturally equipped with this functionality. In XtremWeb, the sandbox acts at two levels. One for the task performer, the worker itself, and a more restrictive one for the downloaded Java byte code.

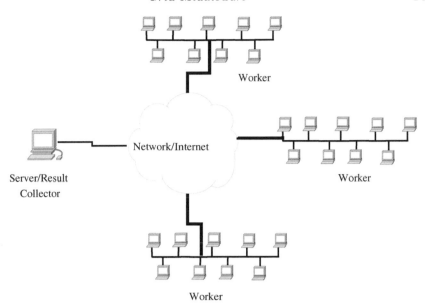

FIGURE 5.9: Architecture of XtremWeb.

On the contrary, binary (or native) applications have complete access to the host system by default. Workers are configured to run any binary task inside a sandbox. These tasks are fully customizable in terms of their memory usage and file system operations.

While security has been handled effectively in XtremWeb, it largely remains a centralized system. XtremWeb-CH is an improved version of XtremWeb, which has a decentralized architecture. It allows execution of high performance applications on a highly heterogeneous distributed environment. In XtremWeb-CH, a worker can directly *talk* to another worker, whereas in XtremWeb, communication exists only between the worker and the server.

5.3.5   Grid Portals

Grid resources encapsulate a lot of hardware and software ranging from high performance computers (HPC) and storage systems to expensive and specific instruments like telescopes and particle accelerators. Grid infrastructure provides a set of common services or tools in order to access and efficiently utilize the distributed, heterogeneous and large-scale grid resources. The current emphasis in grids is towards the Open Grid Service Architecture (OGSA). Grid portals provide an easy way for end users (scientists and engineers, most of whom have no grid knowledge) to access these resources transparently. Grid portals is an easy way to access grids and grid services, and the end users just need to focus on the solution of their domain-specific problem. Web-based grid portals have been proven to be effective for computationally complex scientific applications, as well as in commercial fields. Grid portals can deliver complex grid solutions to end users and can be accessed through a web browser having Internet connectivity. There is no need to download or install specialized software or worry about setting up networks, firewalls and port policies. A lot of grid portal projects have been launched. These include the Gateway project [112], Mississippi Computational Web Portal [113], NPACI HotPage [114], and JiPANG [115]. Numerous solutions were put forward for the challenge of connecting distributed resources. The web technology has evolved very fast in this regard. The combination of web technology and grid technology has accelerated the development of grid portals.

5.3.5.1   Grid Portal Solutions

Grid Portals

Web-based grid portals have been proven to be an efficient way for users to access and utilize grid resources. A grid portal acts as a layer closest to the end users in grid architecture. It hides the grid resources' complexity and the sophisticated grid services and offers a unified web interface to the users. Figure 5.10 shows the view of grid architecture with an emphasis on the grid portal layer and the functionalities provided by the grid portal layer. User management includes the creation of user account, user login and logout, the grant of user rights and the issues related to security. Job submission and monitoring enables users to launch their domain-specific applications and retrieve the execution status during run-time, and it is a preliminary function of the grid portal. Workflow is used for grid worklow applications. It manages the state and persistence information of the components, workflow information delivery, etc. Data management is responsible for downloading and uploading user's data files through the grid portal and securing the data transfer via grid portal tools. Information service provides access to directories and status tools. On the right of Figure 5.10 are the lower-level grid middleware tools, called the *portal backend*. The traditional operations and computations such as application invocation, which involves choosing appropriate grid resources for a job, invocation of applications using GRAM, management of grid resources, users and jobs. These are executed by the grid middleware of the backend.

Security Issues

The defacto standard for grid security is the Grid Security Infrastructure (GSI). GSI is a public key-based X.509 conforming system that relies on trusted third parties for signing user and host certificates. Typical usage models require that each user is assigned a user credential consisting of a public

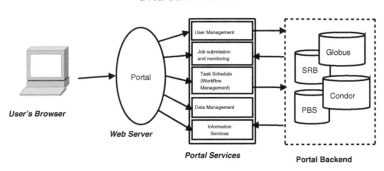

FIGURE 5.10: Grid view with grid portal layer.

and a private key. Users generate delegated proxy certificates with short life spans that get passed from one component to another and form the basis of authentication, access control, and logging. Users can typically manage the credentials (and proxy) manually or use a number of command line-based utilities to manage their credentials. In the grid portal layer, we consider these aspects from the user's point of view. Users never have to know anything about grid security, credentials, proxies or other technical issues. The unique method for users to access grid resources is through the username/password pair. Once logged in, they can use the grid resource through the portal. The management of credentials is handed over to a MyProxy [117] server. The portal backend acquires the user proxy certificate from MyProxy server, which contains a credential repository for the Grid and securely delegates the credentials without transferring the user's secret key through networks.

Portal and Portlet

Portal is a general concept and not a term invented specifically for grid computing. A portal collects diverse grid resources into a web page and displays them with specific features. Instead of separately visiting the disparate sites, users have a unified means for accessing the resources. Portal pages may have different set of portlets, creating content for different users. A portlet is a Java technology-based web component, managed by a portlet container that processes requests and generates dynamic content. Portlets are used by portals as pluggable user interface components that provide a presentation layer to information systems [118]. Java Standardization Request (JSR168) called portlet specification aims to enable interoperability between portlets and portals. JSR168 defines a portlet container that manages portlets. A contract is also defined for the container to call methods during a portlets life cycle. The portlet developer can implement these methods to provide the desired functionality. JSR168 provides the developers one common model that allows them to insert the new functionalities into the portal server. Use of portlet technology to exploit the grid portal can enhance the extensibility and maintainability of the system.

Portlet-Based Web Portal: GridSphere

Grid portal development tools such as Grid Portal Development Kit (GPDK), Grid Portal Toolkit (GPT), and NPACI HotPage are used for the development of grid portals. Although these tools have successfully implemented grid portal functions, they were introduced before the release version of JSR168. As a result, these grid portal development products are unable to unify the portlet concept. A younger grid portal development kit, namely GridSphere, is a portlet based on the web portal toolkit.

GridSphere [116] was developed by the project GridLab. In fact, GridShpere is a common web development environment, and it is not bound together with any Grid technology. It has the objective of constructing a set of well-integrated and easy-to-use portlets. It can be used to realize online collaboration, computation, data management and data display. For this purpose, a high-level API is offered to the users for developing customized portlets.

Grid Portlets is set of portlets cooperating with GridSphere. It offers a number of grid application-oriented portlets, which must be downloaded and deployed into GridSphere. These portlets provide a couple of functions such as credential request and management, searching of resources, job submission and file transfer. Based on these portlets, developers can create more powerful portlets using higher-level APIs.

References

[91] UNICORE. UNICORE objectives. Web Published, 2007. Available online at: `http://www.unicore.eu/unicore/` (accessed January 1st, 2009).

[92] NorduGrid. The NorduGrid collaboration. Web Published, 2007. Available online at: `http://www.nordugrid.org/` (accessed January 1st, 2009).

[93] Netsolve/GridSolve. GridSolve description. Web Published, 2007. Available online at: `http://icl.cs.utk.edu/netsolve/` (accessed January 1st, 2009).

[94] Dorian Arnold, Sudesh Agrawal, Susan Blackford, Jack Dongarra, Michelle Miller, Kiran Sagi, Zhiao Shi, and Sathish Vadhiyar. Users' guide to NetSolve v1.4.1. Technical Report ICL-UT-02-05, University of Tennessee, Innovative Computing Departement, Knoxville, TN, June 2002.

[95] OmniRPC. OmniRPC: a grid RPC system for parallel programming. Web Published, 2007. Available online at: `http://www.omni.hpcc.jp/OmniRPC/` (accessed January 1st, 2009).

[96] Ninf. A global computing infrastructure. Web Published, 2007. Available online at: `http://ninf.apgrid.org/` (accessed January 1st, 2009).

[97] XtremWeb. A global computing experimental platform. Web Published, 2007. Available online at: `http://www.lri.fr/~fedak/XtremWeb/` (accessed January 1st, 2009).

[98] Legion. The Legion project. Web Published, 2007. Available online at: `http://legion.virginia.edu/` (accessed January 1st, 2009).

[99] EGEE. gLite: lightweight middleware for grid computing. Web Published, 2007. Available online at: `http://glite.web.cern.ch/glite/` (accessed January 1st, 2009).

[100] UNICORE Plus. Uniform interface to computing resources. Web Published, 2007. Available online at: `http://www.fz-juelich.de/zam/cooperations/unicoreplus` (accessed January 1st, 2009).

[101] VDT. What is the virtual data toolkit? Web Published, 2007. Available online at: `http://vdt.cs.wisc.edu/` (accessed January 1st, 2009).

[102] GridDaen. Web Published, 2007. Available online at: `http://www.cngrid.org/09_resource/griddaen.htm` (in Chineese) (accessed January 1st, 2009).

[103] Condor. Condor v6.8.5 manual. Web Published, 2007. Available online at: http://www.cs.wisc.edu/condor/manual/v6.8/ (accessed January 1st, 2009).

[104] Douglas Thain, Todd Tannenbaum, and Miron Livny. Distributed computing in practice: the Condor experience. *Concurrency and Computation: Practice and Experience*, 17(2–4):323–356, 2004.

[105] Parvin Asadzadeh, Rajkumar Buyya, Chun Ling Kei, Deepa Nayar, and Srikumar Venugopal. Global grids and software toolkits: a study of four grid middleware technologies. Web Published, 2004. Available online at: http://www.citebase.org/abstract?id=oai:arXiv.org:cs/0407001 (accessed January 1st, 2009).

[106] Andrew S. Grimshaw, Wm. A. Wulf, and the Legion team. The Legion vision of a worldwide virtual computer. *Communications of the ACM*, 40(1):39–45, 1997.

[107] Gregor von Laszewski and Kaizar Amin. *Grid Middleware*, chapter Middleware for Communications, pages 109–130. Wiley, 2004. Available online at: http://www.mcs.anl.gov/~gregor/papers/vonLaszewski--grid-middleware.pdf (accessed January 1st, 2009).

[108] David Snelling. *UNICORE and the open grid services architecture*, pages 701–712. Wiley, 2003.

[109] Andrew S. Grimshaw, Anand Natrajan, Marty A. Humphrey, Michael J. Lewis, Anh Nguyen-Tuong, John F. Karpovich, Mark M. Morgan, and Adam J. Ferrari. *From Legion to Avaki: the persistence of vision*, pages 265–298. Wiley, 2003. Available online at: http://stinet.dtic.mil/cgi-bin/GetTRDoc?AD=ADA447084&Location=U2&doc=GetTRDoc.pdf (accessed January 1st, 2009).

[110] Rajkumar Buyya, Manzur Murshed, David Abramson, and Srikumar Venugopal1. Scheduling parameter sweep applications on global grids: a deadline and budget constrained cost-time optimisation algorithm. Technical Report GRIDS-TR-2003-4, University of Melbourne, Grid Computing and Distributed Systems Laboratory, Australia, June 2003. Available online at: http://citeseer.ist.psu.edu/buyya05scheduling.html (accessed January 1st, 2009).

[111] Rajkumar Buyya, David Abramson, and Jonathan Giddy. An economy driven resource management architecture for global computational power grids. In *PDPTA'2000: Proceedings of the 2000 International Conference on Parallel and Distributed Processing Techniques and Applications, June 26-29*, Las Vegas, USA, 2000. CSREA Press, USA.

[112] Tomasz Haupt, Erol Akarsu, Geoffrey Fox, and Choon-Han Youn. The Gateway system: uniform web based access to remote ressources. In *JAVA'99: Proceedings of the ACM 1999 Conference on Java Grande*, pages 1–7, New York, NY, USA, 1999. ACM Press.

[113] Tomasz Haupt, Purushotham Bangalore, and Gregory Henley. Mississippi computational web portal. *Concurrency and Computation: Practice and Experience*, 14(13–15):1275–1287, January 2003.

[114] Mary Thomas, Steve Mock, and Jay Boisseau. Development of web toolkits for computational science portals: the NPACI HotPage. In *HPDC'00: Proceedings of the 31st IEEE International Symposium on High Performance Distributed Computing*, page 308, Washington, DC, USA, 2000. IEEE Computer Society.

[115] Toyotaro Suzumura, Satoshi Matsuoka, and Hidemoto Nakada. A Jini-based computing portal system. Web Published, 2001. Available online at: `http://ninf.apgrid.org/papers/sc01suzumura/sc2001.pdf` (accessed January 1st, 2009).

[116] Jason Novotny, Michael Russell, and Oliver Wehrens. GridSphere: a portal framework for building collaborations. *Concurrency and Computation: Practice Experience*, 16(5):503–513, 2004.

[117] Jason Novotny, Steven Tuecke, and Von Welch. An online credential repository for the grid: MyProxy. Web Published, 2001. Available online at: `http://www.globus.org/alliance/publications/papers/myproxy.pdf` (accessed January 1st, 2009).

[118] Alejandro Abdelnur and Stefan Hepper. JavaTM Portlet specification. Web Published, 2003. Available online at: `http://jcp.org/en/jsr/detail?id=168` (accessed January 1st, 2009).

Chapter 6

Architectural Overview of Grid Projects

6.1 Introduction of Grid Projects

Worldwide grid projects have contributed to the development of the grid. They have purchased or installed equipment. During these projects, numerous dedicated networks and clusters have been set up, which are currently serving as the basic resources for grid research. During the last decade, grid computing has became a very dynamic field, and many scientific computational fields are now looking for solutions toward grid computing. High energy physics, astronomy, earth observation, and biology problems are now common grid applications. In addition to science and engineering applications, e-business and computational finance have also increasingly appeared as grid applications. These applications have put forward domain-specific requirements, such as massive data processing, multisite geographically distributed databases, high performance computing and high throughput computing. Human resources in grid projects are usually organized into small groups, each group focusing on one particular aspect of the grid technology like security, data management, information and monitoring, scheduling, etc. These groups have developed useful software products and published numerous journal papers and research reports.

In this Chapter, we present an architectural overview of some grid projects including a detailed description of their core technologies, together with a brief comparison of their key features.

6.2 Security in Grid Projects

The basic idea behind the grid is resources sharing. However, this cannot be done without fulfilling certain basic criteria. They are namely the protection of the interest of the resource provider, the assignment of the user rights and the successful execution of the job. The needs of grid security can be summarized in three primary points:

1. Need for secure communication between the grid elements.

2. Need to support security across organizational boundaries.

3. Need for single sign-on for users, i.e., permitting users to use multiple resources or sites.

As security is an important aspect of a grid system, many grid projects have concentrated on this domain. After a general description of grid security in virtual organizations, we discuss the security mechanisms realized in some grid projects.

6.2.1   Security in Virtual Organizations

A virtual organization (VO) is a set of individuals and/or institutions defined by certain sharing rules. In VO, users and resources may be large, unpredictable, and dynamic. Roles of users and resources may also be distinct and dynamic in the sense that not all users share the same rights. In this case, resources of a VO need to be coordinated by the VO in order to work together effectively. Security in VO can be considered as a high-level security. VO's security contains some special security policies. These policies are used for interoperation, mutual authentication and authorization, and resource management with the other VOs. The services dealing with the grid security in a VO system are described as follows:

- *Authentication service*: The authentication service should be agnostic to any specific authentication technology. Resource providers and resource requesters may use different security mechanisms, e.g., Kerberos or PKI. An authentication service is concerned with verifying proof of an asserted identity among different authentication systems.

- *Identity mapping service*: The identity mapping service allows for the transformation of an identity in one identity domain into an identity in another identity domain.

- *Authorization service*: The authorization service resolves the policy-based access control decision. This service receives a credential that represents the identity of a resource requester and authorizes the right and scope of accessing to the resource.

- *VO policy service*: For accessing a particular resource, the resource requester should know about the policy associated with a target resource. In a dynamic grid environment, it is important for resource requesters to discover these policies dynamically and make decisions at runtime. VO policy service manages these policies.

- *Credential conversion service*: The credential conversion service provides credential conversion between different types of credentials.

- *Audit service*: The audit service is policy driven as the identity mapping and authorization services.

- *Profile service*: The profile service is concerned with managing the service requester's preferences and data, which may not be directly consumed by the authorization service. This data might be used by applications that interact with a person.

- *Privacy service*: The privacy service is primarily concerned with the policy driven classification of personally identifiable information. The user can store personal data, identifiable information, etc. For example, in a commercial context, this can be the patient's data.

More details on the security in VO can be obtained in reference [143].

6.2.2   Realization of Security Mechanisms in Grid Projects

In the *EGEE* project, the work group JA3 takes charge of the research of grid security. Security functions proposed by JA3 have been partly implemented as unique authentication to have access to the entire grid, data confidentiality and integrity, fine resource access control (i.e., deny or grant access to a resource for a user, a group of users, a VO, a role) and pseudonymity (i.e., accessing the grid with a pseudonym instead of the real user identity). In addition, JA3 has especially put forward 'data security', which points out the security limitations existing in the EGEE grid. For instance, anybody can get the file list of a storage element without the need of a valid proxy certificate.

In the *LCG* project, the Joint Security Policy Group (JSPG) has designed three security policies: LCG security and availability policy, grid acceptable usage policy, and the virtual organization security policy. Regarding authentication, a repository is designed to contain all accepted CA root certificates. Regarding audit, sufficient data (e.g., log files and process accounting tables) need to be retained from the resource elements in order to trace the process activity resulting from a grid request.

DataGrid's [142] security mechanism is based on Java and WebService. The edg-java-security module consists of two parts: trust manager and authorization framework. Trust manager is an authentication tool, which accepts X.509-based certificates and Globus proxy certificates. Trust manager can deny the certificates listed in the CRL. Using the authorization framework, a coarse-grained authorization can be achieved through the authorization manager. The authorization manager is able to deduce whether a grid client can be associated with a given attribute. In order to protect resources such as

Java servlets and WebService, two filters, namely the authorization filter and the axis handler, are provided. These two modules can block the bad requests and send back to the client information explaining the reasons of failure. Virtual Organization Membership Service (VOMS) manages the information of the user's relationship with the VO. Single sign-on and log-in into multi-VOs are implemented in VOMS. Log-in into multi-VOs means that the user can login into multiple VOs and create an aggregate proxy certificate, which enables the user to access resources in any of them.

OSG's [127] research of security mainly involves an end-to-end auditing system and dynamic host firewall ports. An end-to-end auditing system allows resource owners to specify where auditing information may be generated (either in auditing log files, distributed databases, or somewhere else) and who can have access to the information. The implementation of dynamic firewalls contains some tools and services for dynamically configuring firewalls as well as opening and closing the ports according to the requirements of applications and middleware. With regard to authorization in VO, a subproject named VO Privilege has been designed and implemented with finer-grained authorization so as to improve user account assignment and management at grid sites, and to reduce the associated administrative overhead.

In the *TeraGrid* [138] project, CA management toolkit, gx-map, was developed cooperatively by the TeraGrid project and the NSF Middleware Initiative (NMI) project. In TeraGrid to verify a user's identity, the security group of TeraGrid must adopt photo identification or some other ways. In order to decide whether to trust a given CA, each site must maintain and update the Certificate Revocation List (CRL), which contains credentials that could no longer be used. Globus Toolkit does not provide an automated mechanism to do this. However, with gx-map's functions, users can map their grid credentials (DNs) to their local accounts on each machine, and the user can freely maintain his or her credentials. Gx-map can also operate a set of credentials, which allows automatic update of the CRL.

Grid Operation System (GOS) is grid software developed by the *CNGrid* project. GOS mainly focuses on three aspects of grids: management of grid resources, management of users and grid security. GOS authentication offers the authentication for both user certificate and proxy certificate. GOS has two authorization mechanisms: one is direct access authorization, and the other is token-based authorization. In the latter, the authorization to effectively access the service or virtual service requires getting a token. A token is a piece of XML message written in Security Assertion Markup Language (SAML). The token is issued by the site. GOS also provides an accounting service. Information such as scheduling number, submitting user, begin and end time, current status and statistic data of resource usage are recorded and the amount payable is calculated according to the finished jobs.

The Japanese grid project called *BioGrid* [136] is a data grid that gathers and processes massive databases and datasets on distributed resources. The security research is mainly related to the secure access of a filesystem. BioGrid has combined Grid Security Infrastructure (GSI), developed by Globus, and Self-certifying File System (SFS), which is a secure filesystem developed at the Massachusetts Institute of Technology. BioGrid has created a user-oriented secure filesystem, named GSI-SFS. The objective of GSI-SFS is to provide users with a convenient method for safely sharing data among the storages, each of which is separately located across the untrusted public network. The users may not have the detailed knowledge of the grid. In GSI-SFS, users enjoy the feature of single sign-on. They can transparently access the grid services by entering the user's passphrase. Mutual authentication between a user and a host is performed on demand. In addition, confidentiality and integrity of data are insured by automated encryption and authentication.

In the *DutchGrid* [125] project, the management of CA is the core service. DutchGrid medium-security CA is an internationally recognized certification authority that distributes credentials to users, hosts and services. People can obtain a certificate by completing a form on the web registration page. For example, a medium-security certificate is needed for functions such as accessing grid production infrastructure (LCG, EGEE, etc.), accessing testbeds like EGEE, PPS and P4CTB, registering to a membership of a VO or to other grid authentication purposes.

One of *GridPP* [126] project's products, GridSite, was developed as a grid credential toolkit, and is now an integrable module of the Apache web server. All operations can be carried out in web pages. Its libraries are compatible with CGI programs. These tools can verify different credentials. In particular, it can support GSI proxies and VOMS attribute certificates, as well as the normal X.509 client certificates. GridSite also enables the copying or deleting of files or directories to/from remote servers using `https`.

Summary

Table 6.2 presents a comparison of the security mechanisms in some grid projects.

6.3 Data Management in Grid Projects

Data is the core of the next generation grid computing. In the near future, grid computing will be composed of 10% calculation and 90% data. Obviously,

	Confidential message	Single sign-on	Authentication	Identity mapping	Authorization	VO policy	Audit
EGEE	Data confidentiality and integrity	DONE	Trusted third party (TTP)	Pseudonymity	Fine resource access control; Delegation with proxy certificates; VO membership service		
LCG			Store all accepted CA root certificates in a repository			LCG security and availability policy; Grid acceptable usage policy; Virtual organization security policy	Retain log information from resource elements to trace user's activity
DataGrid		DONE	Accept X.509 based certificate, Globus proxy certificates; Deny the certificates in CRL		Coarse-grained authorization; Authorization filter	Log-in into multi-VOs	

Table 6.1: Comparison of the security mechanisms in some grid projects.

	Confidential message	Single sign-on	Authentication mapping	Identity mapping	Authorization	VO policy	Audit
OSG					Finer-grained authorization		End-to-end auditing
TeraGrid			gx-map: automatically user DN mapping and updating CRL				
CNGrid		DONE	Authentication for both user certificate and proxy certificates		Direct access authorization; Token based authorization		Accounting service
BioGrid	Data encryption	DONE			On-demand authentication		

Table 6.2: Comparison of the security mechanisms in some grid projects (cont.).

data plays a very important role in grid computing. In an increasing number of scientific disciplines, large data collections are emerging as important community resources. In scientific domains as diverse as global climate change, high energy physics, and computational genomic, the volume of interesting data is already measured in terabytes and will soon reach the petabytes. The communities of researchers that need to access and analyze this data, often using sophisticated and computationally expensive techniques, consist of large communities and are almost always geographically distributed. Same is the case with the computing and storage resources that these communities rely upon to store and analyze their data [144].

A data grid is a distributed infrastructure integrating intensive computational capability and large scale databases. Several key technologies in this infrastructure exist, such as data access, data replica management, metadata management, data security, query optimization, realtime data process, etc. Among these technologies, data access, metadata management and data replica management are the core. In data grids, the content of metadata includes information about file instances, content of file instances and various storage systems. In other words, metadata describes the content and structure of data, information content represented by the file, and the circumstances under which the data were obtained. A data grid is based on the elementary functions of the grid. Its core is made up of metadata management and storage resources management. On the one hand, a metadata catalog can efficiently combine heterogeneous resources and services. On the other hand, a storage resource broker can effectively manage different storage resources.

Some of the grid projects mentioned in this chapter emphasize data grid infrastructure. These include European-DataGrid [142], BioGrid [136], EGEE [121], D-Grid [123], and SDG [134]. The details are discussed below.

DataGrid

DataGrid [142], a European project, aims at building the next generation of computing infrastructure, which provides intensive computation and analysis of shared large-scale databases.

DataGrid's Replica Management is middleware enabling the management of files, datasets and collections on the grid. This software, developed within the DataGrid project, is named *Reptor*. Reptor is a virtual single access point for the user to the replica management system. It provides transparent access to the underlying infrastructure. For ubiquitous access and high performance, Reptor provides a SOAP interface. The SOAP interface equips the service with a generic interface. Reptor supports more specialized protocols such as RMI, to achieve the higher performance required. Reptor also provides workflow management functionality for VOs. The response for requests coming from one VO is scheduled according to some predefined priorities so that the available resources can be utilized optimally. Despite Reptor's provision of single point of entry, for fault tolerance reasons Reptor can be configured as a distributed service. In addition, Reptor totally controls the files created or registered through it. Once a file is under Reptor's control, no one else is allowed to delete or to modify the file. Reptor offers commands to register or to unregister a file. File transfer is the core service of Reptor. Its processing API allow pre- and post-processing steps before and after a file transfer. For instance, some compression operations can be placed in these steps. In order to be consistent with the replicas, some consistency services are offered by Reptor:

- *Update propagation* in which once a file is changed the modification should be propagated to all of its replicas.

- *Inconsistency detection*, which can find inconsistencies caused by system failures or attacks.

- *Lifetime management*, some life-limited replica may be created for a temporary utilization of users. Once the predefined lifetime has expired, replicas should be deleted or destroyed.

In Reptor, there are also some other basic services, metadata service, replica selection service based on information coming from network monitoring service, access history service, etc.

Replica Location Service (RLS) has been developed in cooperation with the Globus project. RLS maintains and provides access to mapping information from logical names of the data items to target names. These target names may represent physical location of data items, or an entry in the RLS may be mapped to another level of logical naming for the data item.

Replica Optimization Service (ROS) is based on network monitoring services and the storage element. Networking monitoring service can determine the estimated cost for transferring a file over the grid in terms of bandwidth. The storage element can estimate the cost associated with a file itself, i.e. the time spent for making the data accessible. As of now, optimization has realized a short-term objective: when a job requests a file, the ROS finds the best replica on the grid in terms of the cost required to transfer the file to the local site where the job is running. The long-term objective of ROS is to create and to delete replicas anywhere on the grid according to its predictions of file usage across the grid. For this purpose, an economy-based algorithm is going to be adopted.

Mass storage management includes two potential candidates for the Data-Grid. These are the *Storage Resource Broker (SRB)* and the *Sequential Access to data via Metadata (SAM)* systems. SRB provides a middleware interface between the user and the storage systems that may be distributed at various locations. SRB system is composed of a SRB server, a Metadata CATalog (MCAT), the media drivers that connect physical data and the SRB. The information provided by MCAT is used for authenticating a user. A user's information should be compared with the values stored in MCAT. SAM is a file management system as well as a connector between the storage management system and the data processing layer. In order to optimize the use of storage or other resources, SAM clusters the third-party data storage according to the access patterns. It caches frequently accessed data and organizes requests to minimize tape mounts. It also estimates the required resources before submitting requests and makes administrative decisions concerning data delivery priority with this information. In DataGrid, SRB meets the superficial requirements such as resource discovery through metadata and Globus security and the ability to interface to some existing hierarchical storage managements (HSM). However, the SRB architecture is monolithic and centralized and thus not easily scalable to the distributed requirements of a data grid. SAM has showed some required characteristics, but it doesn't have enough reusable, and interoperable mechanisms. The optimal solutions are still being explored.

EDG Java Security provides secure access to EDG's Java-based web services. It contains two parts: trust manager and authorization framework. Trust manager is in charge of authentication based on an X.509 certificate. Authorization framework defines and implements coarse-grained authorization, integration of authorization functionality and Java servlets or web services and the administrative web interface for the authorization module.

BioGrid

BioGrid [136] project is inspired by 500 databases concerning genome information. These databases are composed of a collection of individuals databases over the Internet. BioGrid has developed a particular infrastructure for sharing information contained in these databases. In the following discussion, we introduce the main contributions of BioGrid.

XML format standardization of data in heterogeneous databases: In heterogeneous databases, different formats are used for storing data; for instance, the same protein could be described in a different way in each of two databases. In order to overcome this complexity, a database conversion system has been developed to transform heterogeneous database formats with different schemes into an XML-based standard format using conversion rules. This XML-based standardization also facilitates the integration of web services. This allows, database services to be accessed by a simple XML-based SOAP protocol.

Automatic update system: Imagine that everyday, 15 million new sequences are added to the databases. Such a large amount of data requires an automatic mechanism for efficient updating operations. Such a system is being developed to perform an efficient all-inclusive sequence search by constructing local copies of public databases and updating them.

Classified sequence for homology search: Biodatabases contain massive redundant data. It is partly caused by the unclassified submissions or submission of data fragments. To solve this problem a mechanism has been introduced to compare the data. Comparison is made among data sequences within a database. A hierarchical classification is dynamically built to minimize the redundance during the research of one sequence.

XML-based ADME information database system: ADME (absorption, distribution, metabolism and excretion of drugs) information is important for designing new drugs. This information is related to other XML-based information, such as those of genes, proteins, and compounds. The associated database system realizes a high-level retrieval, which allows searches based on the relationship between the information.

Coordination with existing web services: In order to be coordinated with the web service, an XML layer has been added between the web service and the grid database. This system can interoperate with existing database services on the web with little reconfiguration.

EGEE

EGEE [121] covers over 90 institutions in 32 countries. Its data storage resources are also considerable; about 5 petabytes (i.e. 5 million of gigabytes) of storage. EGEE project's objective is to manage the massive amount of data generated by LHC. Its data access has a multi tier architecture namely T0, T1, and T2. Users of each tier have different hierarchical right to access the data, so data management is a great challenge in this project. Much software and many services have been designed and implemented to meet this need.

LHC File Catalog (LFC) is a secure file and replica catalog. It maps logical file names to physical replicas of the file. The logical file name is a user-defined alias. Aliases are mutable but they should always be globally unique. LFC supports full portable operation system interface (POSIX) namespace with secure grid access. Both central file catalog and local file catalog modes are supported. Unlike the European DataGrid catalog, LFC has solved most of the performance and scalability problems. It supports cursors for large queries and features such as query requirement timeout and client retry. For a logical file, LFC stores both globally unique identifier (GUID) and logical or physical mappings, which speeds up the operations that contain also these mapping names. Bulk operations are supported. Database transactions and cursors can handle queries of large scale data. Transactions can also cross logical file names (LFN) and storage file names (SFN), which point to the physical locations of file replicas in storage. Kerberos and GSI are adopted as authentication tools in LFC, which allow single sign-on to the catalog user's grid certificates. The client's domain name (DN) is mapped to the uid/gid pair for authorization. At present, LFC server supports Oracle and Mysql as the database backends.

File Transfer Service (FTS) provides a reliable file transfer infrastructure and allows site manager and VO manager to control or manage the file transfer. FTS includes four components: FTS web service, VO agent daemons, channel agent daemons and database. Users submit their jobs through the portal offered by FTS web service. VO agent daemons take charge of assigning work to channel agents according to VO-specific policies. Channel agent daemons execute the file transfers on a given channel. The database records and manages the states of all these components. In summary, in FTS, a channel is a management level concept. A channel may be a point-to-point network connection or various *catch-all* channels, e.g., anywhere to me, or me to anywhere. Once a job is submitted, it is assigned to a suitable channel. These components can be installed on different nodes or in the same node, according to the needs. A practical example of FTS's deployment is the deployment of the CERN FTS Services [145]. In CERN, FTS servers run at T0 to T1 data export. FTS servers at the T1 sites for T1 to T2 and T2 to T1. These servers are also being used for T1 to T1.

Disk Pool Manager (DPM) is a lightweight storage resource manager (SRM) designed to fulfill the storage requirements of T2 sites. The focus is an easy configuration and service management. DPM handles the storage on distributed disk servers. In fact, it handles pools instead of the file system itself. A pool is a group of file systems located on one or more servers. DPM can handle a permanent file system as well as a volatile file system. In a volatile file system, files can be removed by the system at any time, in a permanent file system, files cannot be removed by the system. DPM supports Mysql and Oracle as the database backend.

Hydra provides secure storage of encryption keys. It focuses upon secure storage of the keys and audit. Encrypted files are stored on storage elements. The sensitive information is the encryption key called Hydra Keystore. It is stored in Hydra Service. Keys are distributed to at least three keystores. Even if one keystore doesn't work, this key can be recovered from the other two. One important property is that no key can be reconstructed from the content of one single service. So, if one of the key servers gets compromised, the hacker cannot decrypt the files.

AMGA is a metadata catalog service providing highly available replica of metadata. It is based on RDBMS. It represents a database access service for grid applications. User's or applications can locate their desired data across multiple sites of the grid. Metadata catalog maps a logical filename to the physical location of one or more replicas. The metadata schema can be customized according to the specific user's and application's needs. It also provides a replication layer that offers local availability for the user jobs. AMGA is suitable for the grid environment. Its security mechanism is compliant with grid security. It also hides database heterogeneity.

D-Grid

D-Grid [123] is devoted to the processing of and access to large amounts of scientific data. On this platform, a huge mass of scientific data is collected and shared. These data come from various fields, for instance high-energy physics, astrophysics, medicine, etc. The data management research of this project aims at providing tools for efficient management of these massive data. Three main tools, described in the following paragraphs, are available to meet this objective.

SRM/dCache storage system is a data management system, successfully used in high energy physics for many years. It is also a disk pool management system that maintains data located on distributed nodes. dCache handles all the necessary data transfers between storage elements. In this way it provides users with a transparent file access across different file systems. It can deliver an automatic load balancing for efficient space utilization.

Storage Resource Broker (SRB) presents the user with a single file hierarchy for data distributed across multiple storage systems. It is a middleware layer acting as a uniform interface between application and diverse storage resources. SRB consists of two separate servers, namely the SRB master and the SRB server. The SRB master continuously listens to client requests at a certain port. Once a connection from a client is established and authenticated, it establishes a connection with the SRB server, which we call SRB agent, on behalf of the client. Then, the client and the SRB agent communicate using a different port and the SRB master continues to listen for more connections.

DataFinder is a management tool for large scientific datasets. These datasets can be stored with different protocols or interfaces, such as Web-DAV, FTP, GridFTP, OpenAFS, etc. XML is adopted to describe a data structure and its metadata. All these operations can be managed through the network using the standardized protocol Web-based distributed authoring and versioning (WebDAV). It is a set of extensions to the `http` protocol allowing users to edit and manage files in a cooperative way on remote web servers.

SDG

SDG is based on the scientific data of massive database resources in the Chinese Academy of Science. Scientific grid software is mainly aimed at building a uniform secure system, realizing a unified access interface of distributed and heterogeneous resources. Among these software tools - the Information and Metadata Service (IMS) stores the core catalog of information needed by the Store Service (SS) - Data Access Service (DAS) is the core service, which is established in a secure environment constructed by the security system, and which offers a unified data access interface - SS provides services as, data storage, backup and networking service - Security Service provides other modules with authentication and authorization information, ensuring secure access. These applications are now briefly presented.

Data Access Service (DAS) connects more than 40 research institutes' massive scientific data resources. DAS stands between users and physical databases, and it encapsulates these databases so as to offer a uniform access interface. Presently, DAS supports Oracle 8, Oracle 9i, SQL Server2000 and Mysql 4 databases. Users can access DAS by searching the frontend

DataView. DAS summarizes a layer, named virtual database, from the physical databases. The Web DataView is actually based on the virtual database. To this end, if the database structure is changed, the only thing to do is to modify the configuration, without modification of the program sources.

Information and Metadata System (IMS) provides SDG or other application systems with an information service. IMS stores backend metadata. Currently, it mainly uses Lightweight Directory Access Protocol (LDAP). IMS enables the registration of resources, storage and maintenance of resource information, discovery of resource information, detailed information of data resources and services. IMS includes two main functions: information dissemination and information search. Information dissemination takes charge of the generation, registry, and maintenance mechanisms for resource information. Information research offers searching mechanisms of the resource information.

Store Service (SS) SDG platform is equipped with 20-terabyte disks and 50-terabyte tapes, as well as other software and hardware resources as a super data server. The SSTools realizes the reliable transfer between client-end and server-end. In addition, users can apply disk space through this tool. Briefly, SSTools has implemented three mechanisms: authentication, FTP data transfer and authenticated data transfer, and disk space application and query.

CA System is an important infrastructure for data service activities. PKI is the core of this authentication system, which offers the SDG's digital certificate service. With the certificate issued by CA System, users can access SDG resources in a secure way.

Summary

Table 6.3 presents a comparison of the data management mechanisms in some grid projects.

6.4 Information Services in Grid Projects

The grid monitoring service is based on the so-called resource information. Some information systems devoted to the collection of the resource information are developed or adopted within each grid project. In addition, the information gathering layer is also an important part of the information and monitoring service. Finally, the information representation is well accomplished by graphical modeling tools such as the GLUE schema. Among the following grid projects, some of them have integrated existing tools into their grid service while others have implemented their own tools for monitoring the grid resources. We now investigate these projects in detail.

	Replica management	Data access	Storage management	Data transfer	Metadata management	Data security	File system or database
DataGrid	Reptor provides a single point of entry, supports SOAP	Access history service RLS (Replica Location Service); Replica selection service	SRB (Storage Resource Broker); SAM (Sequential Access to data via Metadata)	Pre- and post-processing steps before and after a file transfer. Consistency service	File metadata, collection metadata, security metadata, application metadata, management metadata	Trust manager and authorization framework, based on X.509 certificate, supporting coarse-grained authorization	File system
EGEE	LFC (LHC file catalog): maps logical file names to physical replicas of the file.		DPM (Disk Pool Manager): lightweight storage resource manager (SRM)	FTS (File Transfer Service): allows manager to control file transfer	AMGA, a metadata catalog service	Hydra: secure storage of encryption keys	File system
D-Grid	DataFinder		SRM/dCache; SRB;				File system

Table 6.3: Data management mechanisms in some grid projects.

	Replica management	Data access	Storage management	Data transfer	Metadata management	Data security	File system or database
BioGrid		XML format standardization of heterogeneous databases; coordinate with web services					Database
SDG		DAS offers a uniform access interface, supports Oracle, SQL Server, Mysql		SS (Store Service): authentication, FTP data transfer and authenticated data transfer, disk space application and query	IMS (Information and Metadata System), uses LDAP; Main functions: Information dissemination and information search		Database

Table 6.4: Data management mechanisms in some grid projects (cont.).

EGEE

In *EGEE* [121], the JRA1 work group has carried out research work on information systems and monitoring. This work is aimed at offering information of the grid itself, and at providing end users with a reliable service discovery mechanism. The JRA1 work group has developed two products, R-GMA and SD, in which R-GMA is a collaboration with the DataGrid project.

Relational GMA (R-GMA) provides services for information, monitoring and logging in a distributed computing environment. It is realized based on the Grid Monitoring Architecture (GMA) as recommended by the Global Grid Forum (GGF). The GMA model considers that a grid information infrastructure consists of *consumer* (requiring information), *producer* (offering information), and *registry* (brokering communication between consumer and producer). R-GMA has combined GMA with relational database query language SQL in order to manage grid information (record, insert and delete). In this way, R-GMA presents to users a virtual database of grid information. The information and monitoring system appears as one large relational database.

gLite Service Discovery (SD) implements service details for access (service name, service type, service version, etc.) by offering a standard interface. Its API mainly has the following functions: getting information about a known service, finding a service. SD service needs the support of the information system because the SD API is a client application for each information system it uses. At present, R-GMA information system, BDII information system and XML FILE are the underlying information systems of the SD.

Compute Element Monitor (CEMon) is responsible for gathering information coming from CE. This information is represented in the GLUE Schema [146], which maps the grid resource information into a concrete schema. In EGEE, CEMon isn't adopted; however, it is currently used by the OSG grid project.

CrossGrid

OCM-G monitoring system is used for monitoring grid applications. It supports both cluster and grid environments based on Globus 2.4. It provides on-line information about a parallel or distributed application, such as CPU usage, delay and volume of communication, to upper layer tools, as performance analysis tools. OCM-G contains three components, Service Manager (SM), Local Monitor (LM) and Application Process (AP). One SM for each site enables routing monitoring requests. One LM for each host/node helps to monitor the ongoing application execution. Thus, a set of LM and SM involved in the monitoring of a single application form a virtual monitoring system.

GÉANT/GÉANT2

GÉANT2 offers a Multi-Domain Monitoring (MDM) service that allows users to access network performance from multiple domains. It can monitor one or several domain networks using standardized interfaces. In order to offer multi-domain-crossed monitoring information, such a system should also talk to each domain using their interfaces. The implementation of MDM is based on the perfSONAR [146] protocol. PerfSONAR is founded on SOAP XML messages and services to communicate with each domain. These services can be the Measurement Point (MP) service, which offers current monitoring data, or the Measurement Archive (MA) service, which offers historical monitoring of data.

OSG Monitoring and Information Services

OSG [127] has adopted multiple grid monitoring systems. Some of them are developed by OSG; the others come from existing products. However, sometimes the information and monitoring tools show different data. OSG is making the effort to find the reasons.

Virtual Organization Resource Selector (VORS) presents the situation of the computational or storage resource in the representation of a map plus data tables displayed on a web page. Resources managed by a given VO may cross multiple grids and countries over the world. VORS uses several different methods to acquire information on the resources that depends on the source of information and the way the information is obtained, for example, withdrawing resource information from OSG registration database; obtaining dynamic test information by cyclically inquiring about the grid sites' specialized web interface GridScan; utilizing information systems, such as BDII, GLUE, GIP, etc.

Generic Information Provider (GIP) is a grid information service. It aggregates static and dynamic resource information. It shows results information in a GLUE schema. It can be dynamically implanted and it provides up-to-date information about the source usage.

GridCat acts as the OSG CE catalog. It is a high level grid cataloging system. The catalog contains information on site with much other valuable information of each site to help job submission and job scheduling for application users and grid scheduler developers.

MonALISA provides a distributed monitoring service system using JINI/Java and WSDL/SOAP technologies. Each MonALISA server acts as a dynamic service system and provides the functionality discovered and used by any other services or clients that require such information.

CNGrid

CNGrid's information and monitoring system is a cooperative service of the following three components. Monitoring Data Provider (MDP) is developed by the CNGrid project. It is running on nodes, and is responsible for gathering the resource data. The data gathering time interval can be assigned through the configuration or deployment of MDP. Each node has a Community Monitor Service (CMS), and this service offers information to users or upper-layer applications. Monitor Data Aggregator (MDA) is running on the core nodes of the grid site. It gathers information coming from all the nodes and saves this information into the database located at the core node.

ChinaGrid

In ChinaGrid the information provider is responsible for the collection of the information from the resource properties. Different providers may collect different kinds of information. The provider manager manages the providers, sets their parameters and receives the dynamic provider's reports, checks this information and reports this data to the resource monitor service. It also answers the queries of the resource monitor service that accepts the query request and returns the result. All three components working together accomplish the resource monitor function.

Summary

Table 6.5 presents a comparison of the grid resource information and monitoring tools.

6.5 Job Scheduling in Grid Projects

Job scheduling in a grid environment is a hot spot of the grid technology research. A grid provides a large variety of complex services, including computational resource services, data management services, etc. The interaction of those services requires an extensible and integrated resource management. Although such a coordinated scheduling of services is currently not readily available, here we point out some of the research on scheduling that has been done or ongoing within the existing grid projects.

DataGrid

The workload management section of DataGrid [142] is aimed to define and implement an architecture for distributed scheduling and resource management in a grid environment. The Workload Management System (WMS) consists of the following components: job description, job submission, job monitoring and control, as well as user credentials information querying. The GUIs for these components are also designed and implemented.

In job description, a JDL editor is available for users. It can parse and

Tool name	Function	Layer level	Used or developed by
R-GMA	Information collection and monitoring + relational DB interface	Information system + information gathering	EGEE
gLite service discovery	Getting information; Finding service	Information gathering	EGEE
CEMon	Gathering information of CE	Information gathering	Developed by EGEE; used by OSG
OCM-G	Providing on-line information about application	Information system (LM) + CrossGrid	
MDM	Multi-domain standardized interface networks monitoring	Information system + information gathering	GÉANT
VORS	Multi-VOs information monitoring	Information gathering	OSG
GIP	Aggregating static and dynamic resource information	Information system	OSG
GridCat	Acting as OSG CE catalog	Information system + information gathering	OSG
MonALISA	Monitoring service using JINI/JAVA and WSDL/SOAP technologies	Information system	OSG

Table 6.5: Grid resource information and monitoring tools.

generate job descriptions in JDL and XML. JDL is used not only to describe jobs but also to describe the resource and data requirements.

Job submission is executed on the remote computing elements. It can restart the job from a previously saved checkpoint and interactively communicate with the running job. In addition, the task of resource discovery is also considered during job submission. Through the job submitter GUI, users can specify a specific network.

Finally, the monitoring and control module can query the information about job status, cancel one or more submitted jobs and retrieve the job result (output files) of the completed job(s).

CrossGrid

CrossGrid Scheduler is also one of the key products of CrossGrid [124]. The job description language can express sequential and parallel jobs. It also supports the permanent job queue and multiple users. The JDL coming from the client is submitted to a scheduling system, which contains three components: scheduling agent, resource selector and application launcher. CrossGrid scheduler has also integrated the DataGrid job submission service.

The scheduling agent accepts the request of JDL from the service interface; the batch and interactive jobs are also accepted. These jobs will be launched in the following steps, according to the job priorities.

The resource selector can choose the resources across multiple sites. It selects the computational resources based on the user's requirements and the job types.

The application launcher submits the sequential or parallel applications to the grid source. The application launcher also offers fault tolerance mechanisms.

GridPP

SAM-Grid is cooperatively developed by the project GridPP. SAM-Grid is based on SAM, a data handling tool, and adds the functionality of grid job scheduling and brokerage. The standard Condor-G is used as the job scheduler so as to enable scheduling of data-intensive jobs with flexible resource description. SAM-Grid has optimized the Condor-G middleware in three areas: automating the selection of grid site on behalf of the user's application; extending the job/resource description language ClassAd with the matchmaking framework to externally include functions to be evaluated at the match time; removing the restriction that the job submission client had to be on the same machine as the queuing system, which enables a multilayered architecture.

EGEE

Computing Resource Execution and Management (CREAM) is a lightweight job management service at the CE level. CREAM accepts the submission described in JDL and supports batch and parallel (MPI) jobs as well as bulk jobs. The other job management functions such as job status retrieval, job listing, and job cancelation are also supported.

gLite Workload Management System (gLite WMS) comprises a set of grid middleware components responsible for the distribution and management of tasks across grid resources. As gLite WMS uses the same JDL as CREAM, we can easily integrate gLite WMS and CREAM together. In fact, CREAM can work as a grid component of gLite WMS. That is to say that jobs submitted to the gLite WMS can be forwarded for their execution on CREAM-based CEs. Besides CREAM in the gLite WMS architecture, there are other important cooperating services as well. Logging and bookkeeping service tracks events generated by different components as a job is being executed. Data management service duplicates and transfers data, and manages data replicas. Information system offers the hardware/software information. At present, gLite takes effort to standardize the interface for every component service with the web service interface for wider utilization.

TeraGrid

TeraGrid has attempted to combine job management service with workflow conception. Thanks to workflow, the computation can be expressed in a form that can be readily mapped to the grid. This idea is elaborated by the adoption of Virtual Data Language (VDL). VDL is explored by the GriPhyN [147] Project. This language allows users to specify the construction of grid workflow to derive data.

The description file of a computation written in VDL is abstracted by a virtual data workflow generator into an abstract workflow and sent to the grid work execution component or worker nodes. In TeraGrid, DAGman (a production quality workflow executor developed as part of the Condor project) acts as this executor. Between these two components, the workflow spec (here VDL) and the executor (here DAGman), there is a component called the planner, which creates the execution plan. In detail, the planner distributes the workflow over different resource sites. TeraGrid has employed Pegasus [148] as the planner. The Pegasus planner converts abstract workflows to concrete workflows and deploys them to the grid resources.

ChinaGrid

In ChinaGrid, a module named execution management accomplishes the function of job management. Execution management is an important function component in the China Grid system platform (CGSP). It enables applications to have coordinated access to the underlying resources. The visible characteristics consist of a uniform interface for job creation, job creation monitoring and monitoring of other services (such as WS, WSRF, JSDL, composite service, etc.), support for legacy binary program execution management, and distributed workflow engine management and workflow balancing. Clearly, there are three logical layers in the architecture of execution management: job submitter, job manager and job executors.

Job executors offer several services for carrying out the execution of the job. Among these services, the most notable one is the General Running Service (GRS), which is a normal WSRF-compliant web service. It can deal with any executable program. GRS controls the execution and scheduling of the tasks at the node level. The goal and motivation of this module is to implement the virtualization of computing resources, to provide a uniform submission interface for legacy program tasks, and to execute legacy programs without any modification in the grid environment. The legacy program means command line programs and MPI-based programs. However, MPI support is still under exploration.

On top of the job executor layer is the job manager who is responsible for distributing the jobs, initiating the execution, and managing and monitoring the execution. A workflow manager is a member component of the job manager. Under this component, workflow is running as a web service in the axis container. The workflow manager supports workflow service deployment, workflow submission, and workflow monitoring and controlling. This workflow manager can balance the possible workload among the workflow engine and dispatch service requests to specific workflow engines (a workflow engine implements basic workflow execution functions such as execution, control, monitoring, etc.).

Summary

Table 6.6 presents a comparison of the components and the characteristics of some job scheduling tools.

6.6   Grid Applications

By integrating distributed computational and storage resources, grid computing offers powerful computing capacities and huge storage spaces. With the ability to build and deploy applications that can take advantage of the distributed resources, grids provide an environment well suited for developing truly innovative and dynamic solutions. With the support of grids and especially with datagrids, the capacity for information sharing and information processing is achieved at an unprecedented level. Furthermore, the research capacities in different fields have also been greatly improved. In this section, we introduce some key application areas of grid computing.

	Components contained	Job description	Accepted job type	Support for workflow
WMS (DataGrid)	Job description; Job submission; Job monitoring and control	JDL able to describe sequential and parallel job		No
CrossGrid scheduler (CrossGrid)	Scheduling agent, resource selector and application launcher	JDL	Batch and interactive job	No
CREAM + gLite WMS (EGEE)	Executor on CE (CREAM); gLite WMS; Other service	JDL	Batch and parallel job (MPI)	No
TeraGrid	Workflow generator; Execution planner; Workflow executor (DAGMan)	VDL		Yes
ChinaGrid	Job submitter; Job manager; Job executor		Any executable program	Yes

Table 6.6: Job scheduling tools composition and characteristics.

6.6.1 Physical Sciences Applications

Physical science is a rapidly growing field along with the development of grid technology. CERN's LHC project launched in 2007 to provide massive data throughput (from 10 Gb/s to 100 Gb/s) where each physics event can be processed independently, resulting in trillion-way parallelism. This great challenge cannot be solved by any single computer or cluster of computers. Hence, scientists are now trying to find ways to solve this problem with new possibilities provided by grid computing. For this purpose, several grid projects are created or involved: LCG, EGEE, DATAGrid of Europe, the UK GridPP and Italy INFN.

6.6.2 Astronomy-Based Applications

Astronomy has always been a dynamic field of research as human beings are always curious about outer space. From processed images of planets to huge amounts of raw data, there is a vast amount of astronomy-based data available on the Internet. This results in large storage requirements. The grid projects D-Grid, DutchGrid of Europe, Chinese OSG, ChinaGrid, American SDG are all involved in this research field.

6.6.3 Biomedical Applications

Computational biology has changed from a traditional computational-intensive science to a high-throughput, data-driven science. Much experimental and measuring equipments is directly connected to computer resources to rapidly obtain and process data. DataGrid technology has also been developed to gather and securely process huge datasets and databases. The Japanese grid project named BioGrid has the objective to facilitate the link between these databases for data processing requiring ultrahigh-speed computing resources. In Europe, the project EGEE has offered grid resources and solutions to distribute data, algorithms, computing and storage resources for genomics. In fact, many countries have established special grid projects for computational biological research, aimed at pharmaceutical development and biodynamic elucidation.

6.6.4 Earth Observation and Climatology

The Earth observation satellites return 100 gigabytes of data images per day. The storage equipment on the ground has already stocked more than 1 000 000 gigabytes of data. These data can be used either to analyze the ozone profiles or to detect oil. The grid allows these data to be easily shared between different 'producers' and 'consumers'. For instance, GridData has realized the storage of data distributed across Europe, and established a testbed specialized for ozone data. Related to climatology, a flood forecasting application is currently being migrated from the CrossGrid test bed to EGEE.

6.6.5   Other Applications

In addition to these key applications mentioned earlier, more and more scientists and researchers are beginning research on grid computing. From academics to industries, grid computing is becoming more and more used among business and sciences. For instance, BeInGrid is a grid project focusing on business issues. Its main applications lie in business experiments on grids. The Chinese SDG [134] project and the Japanese NAREGI [137] project have also carried out nanoscience research within the grid. In summary, Table 6.7 shows the application areas of some grid projects.

	HEP	Biomedical	Astronomy	Earth Observation and Climatology	Business	Nanotechnology
EGEE	Yes	Yes				
D-Grid	Yes		Yes			
BeInGrid					Yes	
CrossGrid	Yes	Yes		Yes		Yes
LCG	Yes					
DutchGrid	Yes		Yes	Yes	Yes	
GridPP	Yes		Yes			
Géant2	Yes	Yes	Yes			
OSG	Yes	Yes	Yes			Yes
DATAGrid	Yes	Yes		Yes		

Table 6.7: Application areas of some grid projects.

References

[119] ChinaGrid. China Education and Scientific Research Grid Project. Web Published, 2007. Available online at: `http://www.chinagrid.edu.cn/chinagrid/index.jsp` (accessed January 1st, 2009).

[120] ChinaGrid. The ChinaGrid project. Web Published, 2007. Available online at: `http://www.chinagrid.net` (in Chineese) (accessed January 1st, 2009).

[121] EGEE. The EGEE project: enabling grids for e-science. Web Published, 2007. Available online at: `http://www.eu-egee.org/` (accessed January 1st, 2009).

[122] Grid'5000. Grid'5000 at a glance. Web Published, 2007. Available online at: `https://www.grid5000.fr/` (accessed January 1st, 2009).

[123] D-Grid. D-Grid initiative. Web Published, 2007. Available online at: `https://www.d-grid.de/` (accessed January 1st, 2009).

[124] CROSSGRID. Key project achievements. Web Published, 2007. Available online at: `http://www.crossgrid.org/main.html` (accessed January 1st, 2009).

[125] DutchGrid. Large-scale distributed computing in the Netherlands. Web Published, 2007. Available online at: `http://www.dutchgrid.nl/` (accessed January 1st, 2009).

[126] GridPP. UK computing for particle physics. Web Published, 2007. Available online at: `http://www.gridpp.ac.uk/` (accessed January 1st, 2009).

[127] Open Science Grid. Science on the Open Science Grid. Web Published, 2007. Available online at: `http://www.opensciencegrid.org/` (accessed January 1st, 2009).

[128] Boinc. Berkeley open infrastructure for network computing. Web Published, 2007. Available online at: `http://boinc.berkeley.edu/` (accessed January 1st, 2009).

[129] GarudaIndia. GARUDA: the national grid computing initiative. Web Published, 2007. Available online at: `http://www.garudaindia.in/` (accessed January 1st, 2009).

[130] EU-IndiaGrid. Joining European and Indian grids for e-science network community. Web Published, 2007. Available online at: `http://www.euindiagrid.eu/` (accessed January 1st, 2009).

[131] LHC Computing Grid Project. Worldwide LHC computing grid: distributed production environment for physics data processing. Web Published, 2007. Available online at: `http://lcg.web.cern.ch/LCG/` (accessed January 1st, 2009).

[132] BeInGrid. Business experiments in grid. Web Published, 2007. Available online at: `http://www.beingrid.com/` (accessed January 1st, 2009).

[133] Globus Alliance. Globus. Web Published, 2007. Available online at: `http://www.globus.org/` (accessed January 1st, 2009).

[134] Scientific Data Grid (SDG). Web Published, 2007. Available online at: `http://www.sdg.ac.cn/` (in Chineese) (accessed January 1st, 2009).

[135] China National Grid. Web Published, 2007. Available online at: `http://www.cngrid.org/en_index.htm` (accessed January 1st, 2009).

[136] BioGrid. Construction of a supercomputer network. Web Published, 2007. Available online at: `http://www.biogrid.jp/` (in Japanese) (accessed January 1st, 2009).

[137] NAREGI. The national research grid initiative. Web Published, 2007. Available online at: `http://www.naregi.org/index_e.html` (accessed January 1st, 2009).

[138] TeraGrid. The TeraGrid project. Web Published, 2007. Available online at: `http://www.teragrid.org/` (accessed January 1st, 2009).

[139] GEANT2. The GEANT2 project. Web Published, 2007. Available online at: `http://www.geant2.net/` (accessed January 1st, 2009).

[140] GEANT. The GEANT project. Web Published, 2007. Available online at: `http://www.geant.net/` (accessed January 1st, 2009).

[141] Infn. The Infn grid project. Web Published, 2007. Available online at: `http://grid.infn.it/` (accessed January 1st, 2009).

[142] DataGrid. The DataGrid project. Web Published, 2007. Available online at: `http://eu-datagrid.web.cern.ch/eu-datagrid/` (accessed January 1st, 2009).

[143] Nataraj Nagaratnam, Philippe Janson, John Dayka, Anthony Nadalin, Frank Siebenlist, Von Welch, Ian Foster, and Steve Tuecke. *The security architecture for open grid services.* GGF OGSA Security Workgroup, 1st edition, July 2002.

[144] Reagan Moore, Chaitanya Baru, Richard Marciano, Arcot Rajasekar, and Michael Wan. The SDSC storage resource broker. In *CASCON'98: Proceedings of the 1998 Conference of the Centre for Advanced Studies on Collaborative research, Toronto, Ontario, Canada, Nov.30-Dec.3,*

1998, The grid: blueprint for a new computing infrastructure, pages 105–129. Morgan Kaufmann publishers Inc, 1998.

[145] CERN.   FTS Services.   Web Published, 2007.   Available online at: `https://twiki.cern.ch/twiki/bin/view/LCG/FtsTier0Deployment` (accessed January 1st, 2009).

[146] perfSONAR. GLUE Schema. Web Published, 2007. Available online at: `http://glueschema.forge.cnaf.infn.it/` (accessed January 1st, 2009).

[147] GriPhyN. Grid physics network. Web Published, 2007. Available online at: `http://www.griphyn.org/` (accessed January 1st, 2009).

[148] Pegasus. Planning for execution in grids. Web Published, 2007. Available online at: `http://pegasus.isi.edu/` (accessed January 1st, 2009).

Chapter 7

Monte Carlo Method

7.1 Introduction

The Monte Carlo method is widely used in many areas of scientific research. From computational physics to fluid dynamics, this method has seen exponential growth with the advent of computationally powerful computers. Indeed, the Monte Carlo method is of great interest for solving systems with unknown analytical solutions. In the real world, more often than not, straightforward analytical solutions are not readily available. Hence, empirical modeling and numerical simulations are much sought to better understand the physical problems involved. While in the past such modeling and numerical simulations were not very accurate, ongoing research and advances in computational power have led to more and more sophisticated and high quality models to better approximate the physical problems. Despite that the computational power has grown exponentially over the years, this power is not able to keep up with the ever increasing demands of the improved model developed by researchers. Hence, the advent of grid systems provides the industry with a powerful tool to tap the resources offered by parallel computer networks. Such networks have theoretically a limitless amount of computational power. So far, there is a great tendency for the industry to adopt grid solutions.

7.2 Fundamentals of the Monte Carlo Method

The Monte Carlo method first saw its application in the computation of the number π. To derive the numerical value of π, one possibility is to calculate numerically the area of a circle A_{circle} of radius r, and then to deduce the value of the number π using the relation $A_{circle} = \pi r^2$. To calculate the area of the circle, we start by filling up a square, of width $2r$ to contain the circle, with N points distributed randomly. At each point, we take note of its position with respect to the circle. If the point falls within the circle, we group it under the set C. The number of points in C is denoted by N_C. Likewise, we group all the points falling outside the circle under the set B. An approximation of the area of the circle can be calculated as the product between the ratio of the number of points in C to the total number of points in $B \cup C$ and the area of the square. The relation is the following:

$$A_{circle} \approx \left(\frac{N_c}{N}\right) A_{square}.$$

This approximation tends towards the exact value of the area of the circle as the number of points N tends to infinity, i.e.,

$$A_{circle} = \lim_{N \to \infty} \left(\frac{N_c}{N}\right) A_{square} = \pi r^2.$$

The value of π is then easily obtained with the relation $\pi = A_{circle}/r^2$.

7.3 Deploying the Monte Carlo Method on Computational Grids

7.3.1 Random Number Generator

As we have seen in Section 7.2, the Monte Carlo method uses the stochastic landing of points in the square to evaluate the area of the circle. This area tends towards the exact value of the area of the circle as the number of points tend towards infinity. However, in order for this method to work, the landing, or in other words the trajectories, of points has to be truly randomized. This is easily seen in the extreme case where the landing of points on the diagram is deterministic and all points land on a single point. In this case, the area cannot be calculated even if the number of points or trajectories of points tends towards infinity.

The key point for the Monte Carlo method is definitely "How to ensure the random property of the trajectories of the points ?" To fulfill this property, we can use the random number generators provided in programs such as MatLab. However, the series of random numbers generated with the software is often only truly random for a relatively small amount of generated random numbers, after which, the generated numbers loop back and the generated numbers can be predicted in a deterministic way. Furthermore, as Monte Carlo computations become more and more complex [150], the amount of required random numbers becomes greater and greater. Thus, standard random number generators with a maximum generation limit of random numbers can no longer fulfill the needs of such computations. In order to circumvent this problem, the use of parallel infrastructures in grid systems has been proposed in reference [149]. This article proposes to generate a sequence of random numbers that stays truly random even when the number of generated numbers is very large.

7.3.2 Sequential Random Number Generator

We now illustrate the method initially proposed in reference [149]. First, a vector S_0 containing the first p random numbers as its vector components is generated: $S_0 = (x_{-p}, x_{-p+1}, \ldots, x_{-p+p})^T$. This is done by using standard random generators built-in with software such as Matlab, Scilab, C++, Fortran, etc. The vector S_0 is also call the *seed vector* because the rest of the sequence of random numbers is generated on the basis of this vector. Secondly, we apply an *additive random matrix* to the seed vector in order to obtain the next vector $(x_{-p+1}, x_{-p+2}, \ldots, x_{-p+p+1})^T$, where the last coefficient, i.e., the random number x_{-p+p+1}, has been generated. The additive random matrix is repeatedly applied to obtain sequentially all the random numbers of the expected sequence.

The additive matrix proposed in reference [149] for instance is a $p \times p$ matrix with integer coefficients M_{ij} having the following properties:

$$M_{ij} = 1 \quad \text{if} \quad j = i + 1$$
$$M_{ij} = 1 \quad \text{if} \quad i = p \quad \text{and} \quad j = 1$$
$$M_{ij} = 1 \quad \text{if} \quad i = p \quad \text{and} \quad j = q$$
$$M_{ij} = 0 \quad \text{otherwise}$$

where q is a parameter given by the user, see reference [149].

7.3.3 Parallel Random Number Generator

The sequential method of generating random numbers can be parallelized and applied in grid networks by breaking the sequence to be generated into different segments where each segment can be generated by a different node located in the grid. For example, in order to generate 10^{11} random numbers, we can divide this sequence of 10^{11} numbers into 10^4 segments of 10^7 numbers. Each node of the grid is responsible for generating 10^7 random numbers based on a different seed vector for each node. These vectors are usually called the *node seed vectors*. We now describe the full parallel algorithm.

1. Generation of the node seed vectors:

   For the first node of the grid, numbered node zero, the node seed vector is obtained by the generation of random numbers using standard random number generator. This is the seed vector $S_0 = (x_{-p}, x_{-p+1}, \ldots, x_{-p+p})^T$ mentioned earlier.

   For the second node, numbered node one, the node seed vector is obtained in two steps. First, the seed vector S_0 generated by the node number zero, is copied and stored locally. The additive matrix M is then raised to the power of $1 * 10^7$ to obtain the matrix M_1. Secondly, the matrix M_1 is applied to the seed vector S_0 to generate the node seed vector $S_1 = (x_{10^7-p}, x_{10^7-p+1}, \ldots, x_{10^7})^T$. This is also called the node seed vector for node one.

For the other nodes, numbered node n, the matrix M_n is obtained by raising the matrix M to the power of $n * 10^7$ and the node seed vector for node n denoted S_n is generated by applying M_n to S_0.

2. Generation of the sequence of random numbers:

   The matrix M is then applied repeatedly to each node seed vector to generate the random numbers in the sequence.

In this way, the node seed vector for each node is generated in parallel independently. Furthermore each node performs in parallel the generation of its own set of random numbers.

This brings us to a practical aspect of the implementation. As M is a $p \times p$ matrix, raising this matrix to the power of $n * 10^7$ proves to be very costly in terms of computational time for large values of n and/or p. As the matrix M is highly sparse, the algorithms for such matrix computation can be optimized to reduce the computational time to construct the matrix $M_n = M^{n*10^7}$. How this algorithm can be optimized is beyond the scope of this book and the reader is referred to [159]. Nonetheless, if the matrix computation of M_n is optimized, the parallel algorithm described previously proves to be more robust and stable for large random number generation.

7.3.4 Parallel Computation of Trajectories

In the previous section, we discussed the use of parallel characteristics of grid infrastructure to generate random numbers. Once the sequence of random numbers is obtained, the stochastic aspect of the trajectories is achieved and the simulation of the Monte Carlo trajectories can begin.

As the simulation of each trajectory is independent of the others, they can also be parallelized using the grid infrastructure. In this case, the implementation is simpler than that of the parallel random generator as the number of operations to be performed is the same on each node. On the contrary, the complexity of the random generation increases with the node number n due to the construction of the matrix M_n. With the sequence of generated random numbers, each node in the grid can take up a segment of this sequence to simulate a series of trajectories. For instance, if each Monte Carlo trajectory requires 10^3 random numbers, each node can simulate 10^4 trajectories by using a total of 10^7 random numbers from the sequence of generated random numbers. A concrete example is now presented.

7.4 Application to Options Pricing in Computational Finance

7.4.1 Motivation of the Monte Carlo Method

How is Monte Carlo used in finance?

We now illustrate how the Monte Carlo method can be used in computational finance, notably in the pricing of derivatives products such as options, futures and forward contracts. In computational finance, using the Monte Carlo method requires the assumption that the movement of stock prices follows a stochastic diffusion process. This allows us to simulate a trajectory of the evolution of the stock prices upon time (hours, days, year, etc.). This simulation is then performed for N trajectories in order to get an approximate distribution of the stock prices at a certain expiry date T. This approximate distribution tends towards the normal distribution as N tends towards infinity.

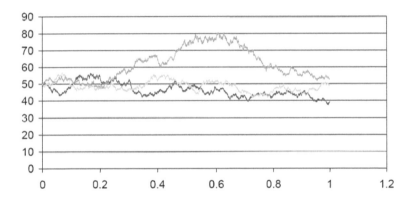

FIGURE 7.1: Simulation of stock prices. Illustration of three trajectories for an expiry date of one year.

In Figure 7.1, we start with a stock price S_0 equal to \$50 and we simulate the evolution of the stock price within an expiry period of one year. Each line in the diagram constitutes an independent trajectory in the Monte Carlo simulation. Each trajectory is evaluated based on the sequence of the random numbers generated as described in Section 7.3.1. Using the stock price values obtained by all the trajectories at the expiry date T gives the expected value $E(X)$ and the possible stock price at the expiry date T. This could be done under the assumption that the distribution of the stock price at the expiry date T is a normal distribution. This assumption is valid according to the Central Limit Theorem as N tends towards infinity. For this reason, it is common to generate a distribution of stock prices at the expiry date by using more than 10^4 trajectories.

With the expected value of the stock prices at the expiry date, we are able to calculate the price of the option at the initial time t_0 based on risk-free evaluation of the market and the possible price of the option at the expiry date [157]. The fundamental notions of risk-neutral evaluation, stochastic processes, log normal distribution of stock prices are going to be discussed in the following sections.

In summary, the price of the option can be evaluated with a certain amount of accuracy that is determined by the standard deviation of the distribution sampled. The illustration above deals basically with path-independent derivative products such as futures and European options.

We now briefly discuss derivative products like American and Asian options that are path dependent; see references [150, 154] for more details. In the case of American options, the evolution path of the stock prices have to be taken into account in order to determine the optimal time to exercise the option. If a similar method is used to evaluate American options, it will eventually lead to a case of "Monte Carlo on Monte Carlo", which becomes computationally impossible to implement on standard computers; see reference [150]. Thus, to price such an option, alternative ways must be adopted, which include binomial trees, trinomial trees and least square method with quasi-Monte Carlo method [158, 156]. Furthermore, the computation of optimal exercising frontier with various American options in a portfolio and the evaluation of the option thereafter allows for the use of the Monte Carlo method.

As a conclusion, pricing methods for European, American and Asian options are well suited for distributed and parallel computing in grid systems due to the independence of data and simulation. In the following paragraphs, we describe the fundamental concepts of these methods.

Monte Carlo Method and Grid Computing

Grid computing is inherently suitable for the Monte Carlo method due to the independence of the trajectories. To better understand why this property is fundamental for implementation in the grid, it is important to understand that the architecture of grid networks is intrinsically parallel. In addition, the nodes, i.e., the computers of the grid are most of the time located in different countries. All these characteristics make grid systems well suited for distributed computations on the different nodes with little communication between these nodes.

During the evaluation of the trajectories of the Monte Carlo method, there is almost no communication involved between the nodes running in parallel. Indeed, if communications were required between the nodes, the communication overhead might be too substantial to make grids a better alternative to standard computation (see reference [149]). Furthermore, for the Monte Carlo method to work, it is advisable that a large value of the number of trajectories be evaluated. A typical value is of the scale of 10^8 or more. If these trajectories are allocated to 10^4 different nodes in the grid, the communication overhead generated between these nodes will be significant and it is likely that this will become the limiting factor for the overall computational speed. Fortunately in the Monte Carlo method the communications appear only after the simulation of the trajectories in order to evaluate the possible price at the expiry date.

Regarding the simulation of the trajectory, the time scale is divided into discrete time steps. These steps are quite small in order to ensure the accuracy of the prices simulated. As the time step decreases, the number of discrete points per simulation path increases. A typical value of the number of discrete points per simulation path is 10^3. Thus, a large quantity of points have to be taken into account in order to implement the Monte Carlo method. Moreover, the trajectory values at two discrete points are computed through the stochastic diffusion process using one random number issued from the sequence generated earlier. This means that very many random numbers and simulations have to be carried out in order for the Monte Carlo method to be accurate. This, in turn, makes the utilization of computational grids interesting as the simulation time can be significantly reduced by using the parallel structure of grids.

If we consider the implementation of the Monte Carlo method, we need a total of $10^8 * 10^3 = 10^{11}$ (total number of trajectories * number of points per path) discrete points to simulate all the possible trajectories. This means that we need to generate 10^{11} corresponding random numbers in order to evaluate the trajectories' point values. On a grid with 10^4 nodes, the generation of random numbers can be divided among the 10^4 nodes with each node generating 10^7 random numbers. This is possible on each node as compared to generating 10^{11} random numbers on a single computer of which the computational time is too long to be of interest. Once the numbers are generated, the simulations of the trajectories can start. In our example, we need to simulate 10^8 trajectories; these trajectories are divided among the 10^4 nodes to simulate 10^4 trajectories for each node. These simulations are carried out and the results on each node are gathered to produce the final result.

Interest of Grids for Financial Institutions

Financial institutions have always been a big investor in information technology with their ever-constant need to automate and speed day-to-day trading operations. This need is also found in middle-office applications where risk management simulations are run to ensure that the institutions are not subjected to too much risk. These simulations usually take up a significant amount of time as almost all possible scenarios are run to ensure full coverage of the risks involved. Grid computing is seeing its application to diverse areas like pricing exotic options, wealth management and trading analytic.

While grid services offer better accuracy and rapidity in results, they are still slow in becoming the de facto standard of financial institutions. Such a slow adoption is due to reasons such as the lack of standards, the lack of reliability and the lack of demonstrated case studies on the utilization of grids tailored for the financial services.

Nonetheless, a recent adoption of grid technology has been launched by Microsoft®. This is a major move in the information technology industry to grid services in various areas, notably in the financial domain. In the past, there have only been several Linux implementations and now we are seeing a major move by Microsoft® to consider grid services in its information technology solutions. This brings about serious orientation towards grid solutions as Microsoft Windows® users still make up a majority of the market. With other major players such as Sun Microsystems® and Exane® contemplating on also having a grid button in its solutions, we do not have to wait too long before grid services enter the financial industry in a big way. In addition, as the adoption of grids starts to gain critical mass, we expect a standard to be created for grid implementation in finance.

7.4.2 Financial Engineering Based on the Monte Carlo Method

In this section, we take a closer look on some basic financial and mathematical notions for the pricing of options using the Monte Carlo method.

European Options

Options are derivative products that allow the owner of the option to buy or sell a certain amount of stock. Contrary to forward contracts, the options owner has the right to sell or buy the stock if he or she wants to. The owner has no obligation to exercise the option. In exchange, for the person who owns the option, he or she has to pay a price to buy the option.

European options are a simple option where option owners can exercise the option only at a predetermined price of the stocks at $t = 0$, named the Strike Price K, and on a predetermined date, named Exercise Date T. In this chapter, we evaluate the price of European options at time $t = 0$ through the use of the Monte Carlo method implemented on the grid architecture. With this simulation, the user knows if it is interesting or not to buy the option at time $t = 0$.

However, before we go into the details of the implementation, let's take a look at some basics to help us to understand the implementation in a better way.

Risk Neutral Evaluation

We assume that the evaluation of the options is done in a risk-free world (see reference [157]). In this world, the returns by investing the money in stocks are the same as that obtained by placing the money in a bank as the investors will not gain extra profits by taking extra risks in the stock market (no risk in risk-free world). In other words, there is no difference in the money returns between the banks and the stock market if it is a risk-free world. Thus, the gains in stocks after a certain time Δt can be obtained by using the interest rates (it is assumed to be risk neutral if money is placed in banks or US treasury bills). For example, if an investor has an amount of cash S_0 at time $t = 0$, the investor is sure to have $S_0 e^{r \Delta t}$ after time Δt without any investment risks involved. The quantity r denotes the interest rate of the bank. We neglect the inflation rate here.

In the case of a European call option, the owner exercises the option at time $t = T$ only if the price of the stock is higher than that of the strike price K and will not exercise it if the stock price is lower than that of the strike price K. Thus, the value of the option at time T will be $\max(S_T - K, 0)$ where K is the price of the stock determined at time $t = 0$ and S_T is the price of the stock at time $t = T$. Based on this price at time $t = T$, we can then calculate the price of the option at time $t = 0$ by discounting the option value at time $t = T$ with the risk neutral interest rate. We obtain

$$C_0 = C_T e^{r(T-0)}$$

where C_0 and C_T are the prices of the option at time $t = 0$ and time $t = T$ respectively.

We are now left with the evaluation of S_T. If we know S_T, we are able to determine C_T and the evaluation of the price of option at time $t = 0$ would be mandatory. However, how can we evaluate the stock price at time $t = T$? The evolution of stock prices over time is nondeterministic. Such random behavior can be modeled by a stochastic process, which is discussed in the next section.

Stochastic Process

The most widely used model for the evolution of stock prices over time is the following:

$$dS = \mu S dt + \sigma S dz$$

where μ is the expected rate of return and σ is the volatility of the stock prices. The variable z is assumed to be a standard Brownian process where $dz = \epsilon \sqrt{dt}$ and ϵ follows a standardized normal distribution, $N(0, 1)$; see reference [152] for more details. The expected rate of return and the volatility are assumed to be constant in the time step dt. In the risk neutral world, the interest rate is taken as the expected rate of return. Thus, based on the price at time $t = 0$, S_0, we are able to simulate a price $S_0 + dS$ at time $t = 0 + dt$

using the random process described above. This procedure is repeated until $t = T$ where we obtain a simulated price S_T at time $t = T$. Such a simulation is a trajectory in the Monte Carlo method.

Evaluation of Stock Prices and Monte Carlo

Based on the simulation procedure presented in the previous section, we are able to simulate many trajectories, which gives us a distribution of the stock price, S_T at time $t = T$. Based on this distribution of S_T, we can then calculate the expected stock price at time $t = T$ by taking the arithmetic mean of this distribution data. This is taken as the stock price, S_T, at time $t = T$.

However, for this method to be valid, many trajectories have to be simulated and this is where grid computations become useful. As each trajectory is independent of each other, the total number of trajectories to be simulated can be carried out on different nodes in the grid. For example, if 10^8 trajectories are required for the distribution to be valid, it will take too long for the simulations to complete on a single computer. However, if these trajectories are distributed among 10^4 computational nodes on a grid, each node needs to compute only 10^4 trajectories of which the computational time is comparably less.

7.4.3   Gridifying the Monte Carlo Method

In this section, we give a concrete example on how grid systems can be used for the pricing of a European option. The proposed programs are written in C++ based on the *gLite* middleware and the *Globus* middleware. The comparison between a sequential and a parallel implementation is made to illustrate the benefits offered by the parallel structure of grids.

As mentioned in Section 7.4.1 under the paragraph entitled *"How is Monte Carlo used in finance?"*, the randomness of the generated random numbers is mandatory for the correct execution of the Monte Carlo method. We now demonstrate how the *gLite* middleware can be used for the generation of such random numbers. Both a sequential and a parallel version are provided below.

Sequential Generation of Random Numbers with gLite

The algorithm for the sequential generation of the random numbers consists of the following steps: get the parameters pertaining to the node, generate the seed vector and finally generate the sequence of random numbers. The program used for the sequential generation of random numbers is presented in Listing 7.1.

Listing 7.1: Sequential generation of random numbers with gLite

```
1   #include <iostream>
```

```
 2   #include <stdlib.h>
 3   #include <cstdlib>
 4   #include <ctime>
 5   #include <fstream>
 6   #include <time.h>
 7   #include "matrix.h"
 8   using namespace std;
 9   #define NUMBEROFRANDNUM 10000
10
11   int main(int argc, char* argv[]) {
12       // --- Get parameters pertaining to the node ---
13       char* inputarg = argv[1];
14       fstream in(inputarg,ios::in);
15       string outrandnumS, outmatrixS, outtimeS;
16       const char* outrandnum, outmatrix, outtime;
17
18       in >> outrandnumS;
19       in >> outmatrixS;
20       in >> outtimeS;
21       in.close();
22
23       outrandnum = outrandnumS.c_str();
24       outmatrix = outmatrixS.c_str();
25       outtime = outtimeS.c_str();
26
27       // --- Generate the seed vector ---
28       srand((unsigned)time(0));
29       vector seed;
30       for (int i=0; i<P; i++)
31           seed[i] = rand();
32       seed.writetofile("",outrandnum);
33       m.writetofile(outmatrix);
34
35       // --- Generate the sequence of random numbers ---
36       vector v;
37       v = seed;
38       for (int i=1; i<=NUMBEROFRANDNUM; i++){
39           if (i%P == 0)
40               v.writetofile("append",outrandnum);
41           v = m1(v);
42       }
43
44       return 0;
45   }
```

In this program, two classes have been defined. They are namely the **vector** class and the **matrix** class. All basic operations such as the addition, subtraction and displaying of a vector have been defined in the **vector** class. The **matrix** class is made up of an array of vectors. Likewise, basic operations such as matrix-vector product have been defined in the class.

In lines 12–26, we get the parameters pertaining to the node. As a result, the variable **outrandnum** contains the name of the file for storing the generated random numbers. In addition, the variable **outtime** contains the name of the file for storing the additive matrix. This variable is mainly for troubleshooting.

In lines 27–34, we generate the seed vector and the additive matrix is prepared. The seed vector and the additive matrix are written to files.

In lines 35–43, we generate the sequence of random numbers and store them to a file.

The output file names are passed as parameters to the workload manager in *gLite* through the job description file. As we can see from the job description file presented in Listing 7.2, the input files are passed as parameters to the compiled program associated to `random_sequential_glite.cpp` during the submission of the job to *gLite*. The results of the job is then recovered from the files `randnum1.txt`, `matrix1.txt` and `time1.txt`. The details on how to submit and collect sequential jobs are explained in Appendix C.

Listing 7.2: Job description file

```
[
  Type = "job";
  JobType ="normal";
  RetryCount =0;
  ShallowRetryCount=3;
  Executable = "random_sequential_glite";
  Arguments = "D1";
  StdOutput = "std.out";
  StdError = "std.err";
  InputSandbox={"random_sequential_glite","D1"};
  OutputSandbox={"std.out","std.err","matrix1.txt",
                 "randnum1.txt","time1.txt"};
]
```

The input file `D1` has the following input parameters:

```
randnum1.txt matrix1.txt time1.txt
```

`randnum1.txt` contains the sequence of random numbers generated through repeated application of the additive matrix. Likewise, the files `matrix1.txt` and `time1.txt` contain the additive matrix and the time to generate the random numbers.

Parallel Generation of Random Numbers with gLite

We now demonstrate how the parallel generation of random numbers can be implemented on *gLite* through the use of C++ programs. The program used for the generation of random numbers is the following:

Listing 7.3: Parallel generation of random numbers with gLite

```
1  #include <iostream>
2  #include <stdlib.h>
3  #include <cstdlib>
4  #include <ctime>
5  #include <fstream>
6  #include <time.h>
7  #include "matrix.h"
8  using namespace std;
```

```
 9   #define NUMBEROFRANDNUM 10000
10   #define NUMBEROFNODES 10
11
12   int main(int argc, char* argv[]) {
13     // --- Get parameters pertaining to the node ---
14     char* inputarg = argv[1];
15     fstream in(inputarg,ios::in);
16     int thisnode;
17     string outrandnumS, outmatrixS, outtimeS;
18     const char* outrandnum, outmatrix, outtime;
19
20     in >> thisnode;
21     in >> outrandnumS;
22     in >> outmatrixS;
23     in >> outtimeS;
24     in.close();
25
26     outrandnum = outrandnumS.c_str();
27     outmatrix = outmatrixS.c_str();
28     outtime = outtimeS.c_str();
29
30     // --- Prepare the additive matrix ---
31     int NumPerNode;
32     NumPerNode = NUMBEROFRANDNUM/NUMBEROFNODES;
33
34     matrix m1(1); // additive matrix
35     matrix m(0);
36     m = m1^(thisnode*NumPerNode);
37
38     // --- Generate the seed vector ---
39     srand((unsigned)time(0));
40     vector seed;
41     for (int i=0; i<P; i++)
42       seed[i] = rand();
43     seed.writetofile("",outrandnum);
44     m.writetofile(outmatrix);
45
46     // --- Generate the node seed vector ---
47     vector v;
48     v = m(seed);
49
50     // --- Generate the sequence of random numbers ---
51     for (int i=1; i<=NumPerNode; i++){
52       if (i%P == 0)
53         v.writetofile("append",outrandnum);
54       v = m1(v);
55     }
56
57     return 0;
58   }
```

In lines 13–29, we get the parameters pertaining to the node. The variable thisnode contains the identity of the node in the grid, i.e., $(1,2,3,\dots)$. The variable outrandnum contains the file name for storing the generated random numbers. The variable outtime contains the file name for storing the additive matrix; this is mainly used for troubleshooting.

In lines 30–37, the additive matrix based on the identity of the node is prepared. This matrix is used to prepare the node seed vector. For the node zero, the additive matrix is kept unchanged. For the nodes 1,2,3,4,..., the additive matrix is raised to the power of `thisnode*NumPerNode`.

In lines 38–45, we generate the seed vector and the additive matrix. Both are saved in files.

In lines 46–49, the node seed vector is generated.

In lines 50–56, the sequence of random numbers is generated and saved to a file.

During the parallel implementation, all the nodes in *gLite* are given the same program. The only thing that differs is the input file given to each node. Based on the information provided to each node via this input file, the identity of each node is dynamically determined during the execution. The execution is then carried out based on the given information. A typical input file looks like:

```
0 randnum1.txt matrix1.txt time1.txt
```

The name of this input file is `D1`. This is the input file for the first node. Other input files for other nodes follow the sequence of file names `D2,D3,D4,` .... In `D1`, the first number is the identity of the node. Here, the identity of the node is 0. The second, third and fourth strings in the file are the file names used for storing the random numbers, the additive matrix and the time required to generate the random numbers. All this information is used as input parameters for the nodes.

During the submission of the parallel job to *gLite*, the input files together with the program are copied and passed to the workload manager in *gLite*. Through the definition of the job by the job description language, the parallel job is submitted to *gLite*.

```
[
    JobType ="Parametric ";
    Executable = "random_parallel_glite ";
    Arguments ="D_PARAM_ ";
    StdOutput = "std_PARAM_ .out ";
    StdError = "std_PARAM_ .err ";
    Parameters =10;
    ParameterStart =1;
    ParameterStep =1;
    InputSandbox ={"random_parallel_glite ","D_PARAM_ "};
    OutputSandbox ={"std_PARAM_ .out ","std_PARAM_ .err ","
        ...randnum_PARAM_ .txt ","matrix .txt ","time .txt "};
]
```

As in the sequential case, the input files are passed as parameters to the copies of the compiled program associated to `random_parallel_glite.cpp` during the submission of the job to *gLite*. The transmission of the parameters is done using the job description file described above. The results of the job are then recovered from the files `randnum_PARAM_.txt`, `matrix.txt` and `time.txt`.

The details on how to submit and collect parallel jobs are described in Appendix C.

Regarding the Globus implementation of this algorithm, the C++ program has to be written with the Message Passing Interface (MPI) library. As mentioned in earlier chapters, Globus works essentially on MPI for parallel computation in computer clusters. The use of *Globus* is to add an additional layer so that a standard communication interface can be established among the clusters. With a common communication interface, clusters based on different platforms can be connected under the same grid. On top of that, the *Globus* grid structure requires a scheduler to eventually dispatch the jobs to the different heterogeneous clusters based on the grid resource information obtained from the grid components in different computer clusters. Hence, such an implementation on MPI/C++ can be migrated on *Globus* without any changes as *Globus* uses the same MPI structure to run parallel computations.

Parallel Simulation of Trajectories with gLite

With the sequence of random numbers generated and stored in the file, we can then proceed with the simulation of the Monte Carlo trajectories using *gLite*. Each node is given a part of the sequence of random numbers generated. This is done by reading the input file whose file name is defined in the E_PARAM_. The variable _PARAM_ identifies the node and takes the value of 1,2,3,.... Below is an implementation using a C++ program on *gLite*.

Listing 7.4: Parallel simulation of trajectories with gLite

```cpp
#include <iostream>
#include <stdlib.h>
#include <cstdlib>
#include <ctime>
#include <fstream>
#include "matrix.h"
using namespace std;

// --- Define constants for European option pricing ---
#define NUMBER_OF_RANDNUM 6000000
#define NUMBER_OF_SIMULATIONS 10
#define TIME_STEP 1 // in days
#define EXPT_GW_RATE 0.0450 //Risk free interest rate
#define VOLATILITY 0.75
#define START_PRICE 1400
#define STRIKE_PRICE 1350
#define EXPIRE_DATE 365 // in days
#define START_DATE 0

int main(int argc, char *argv[]) {
    // --- Get parameters pertaining to the node ---
    char* rand_num_file, rand_gauss_file, option_price_file;
    char* inputarg = argv[1];
    fstream in(inputarg, ios::in);
```

```
25   string rand_num_fileS , rand_gauss_fileS , option_price_fileS;
26
27   in >> rand_num_fileS;
28   in >> rand_gauss_fileS;
29   in >> option_price_fileS;
30   in.close ();
31
32   rand_num_file = rand_num_fileS.c_str ();
33   rand_gauss_file = rand_gauss_fileS.c_str ();
34   option_price_file = option_price_fileS.c_str ();
35
36   // --- Convert random numbers to standard normal random ---
37   double rdn1 , rdn2 , rdngauss;
38   ifstream infile;
39   ofstream outfile;
40
41   infile.open(rand_num_file );
42   outfile.open(rand_gauss_file );
43   for (int i=0; i<P; i++)
44     infile >> rdn1;
45   for (int i=1; i<=NUMBER_OF_RANDNUM; i++){
46     infile >> rdn1;
47     infile >> rdn2;
48     rdngauss = sqrt(-2*log(rdn1))*cos(2*M_PI*rdn2);
49     outfile << rdngauss<<" ";
50     if (i%P==0)
51       outfile << "\n";
52   }
53   infile.close ();
54   outfile.close ();
55
56   // --- Simulate Monte Carlo trajectories ---
57   int duration = EXPIRE_DATE - START_DATE;
58   double Num_of_steps = double(duration )/TIME_STEP;
59   double R_V_gaussien;
60   double stockprices [int(Num_of_steps)+1];
61   double time_step; // in years
62   double options[NUMBER_OF_SIMULATIONS];
63   ifstream infile_gauss;
64   infile_gauss.open(rand_gauss_file );
65
66   for (int j=0; j<NUMBER_OF_SIMULATIONS; j++){
67     double time = 0;
68     stockprices [0] = START_PRICE;
69     time_step = double(TIME_STEP )/365;
70     for (int i=1; i<=Num_of_steps; i++){
71       infile_gauss >> R_V_gaussien;
72       stockprices [i] = stockprices [i-1]*
73         exp((EXPT_GW_RATE-pow(VOLATILITY ,2)/2)*time_step
74           +VOLATILITY *R_V_gaussien*sqrt(time_step ));
75       time = time+time_step;
76     }
77
78     /* Write options total price to file */
79     double stockprice = stockprices [int(Num_of_steps)];
80     if (stockprices [int(Num_of_steps)] >= STRIKE_PRICE)
```

```
81          options[j] = stockprice-STRIKE_PRICE;
82       else
83          options[j] = 0;
84    }
85    infile_gauss.close();
86    double total = 0;
87    for (int i=0; i<NUMBER_OF_SIMULATIONS; i++)
88       total = total+options[i];
89    ofstream out_options;
90    out_options.open(option_price_file);
91    out_options << total;
92    out_options.close();
93
94    return 0;
95  }
```

In lines 9–19, we define constants needed for European option pricing. The variable NUMBER_OF_RANDNUM contains the number of generated random numbers used by the node. The variable NUMBER_OF_SIMULATIONS contains the number of trajectories to be simulated per node. The variable TIME_STEP defines the discrete time step in days. The variable EXPT_GW_RATE contains the risk-free interest rate. The variable VOLATILITY defines the volatility of the stock market. The variable START_PRICE defines the start price of the stock. The variable STRIKE_PRICE contains the strike price of the option. The variable EXPIRE_DATE defines the end date of the simulation. The variable START_DATE contains the starting date of the simulation.

In lines 21–35, we get the parameters pertaining to the node. The input file D_PARAM_ is used to pass different parameters to the program for processing. The variable rand_num_file contains the file name of the file that stores the generated random numbers. The variable rand_gauss_file contains the file name of the file that saves standard normal random variables. The variable option_price_file contains the file name of the file that stores the total option price for all the simulations in the node.

In lines 36–55, we convert random numbers to standard normal random variables. It is important to note that the number of random numbers generated is twice the number of normal random variables needed as we need two uniform random numbers to produce a normal random variable. The variable NUMBER_OF_RANDNUM refers to the number of normal random variables.

In lines 56–77, we simulate the Monte Carlo trajectories. The stock price at the next time step is calculated based on the stock price at the present time step.

In lines 78–93, the options total price is written to a file.

Once the list of random numbers has been generated, the sequence of random numbers is divided among the processors and individual sections of the sequence pertaining to the node are written to the file. The name of this file is stored in E_PARAM_. A typical file of E_PARAM_ has the following contents.

```
rand_num1.txt rand_gauss1.txt option_price1.txt
```

The above is the contents of the file E1. In this case, the sequence of random numbers to be used by the node is stored in rand_num1.txt. The files rand_gauss1.txt and option_price1.txt are used for storing standard normal random variables and the total option price of all the simulations performed by the node. The total option price computed by each node will then be collected to calculate the price of the option.

All the nodes in *gLite* are given the same program and the simulations will be done based on information written in the input files E_PARAM_.

Parallel Simulation of Trajectories with Globus

We now illustrate an example on *Globus* using MPI/C++. Below is a possible implementation.

Listing 7.5: Parallel simulation of trajectories with Globus

```
1   #include <iostream>
2   #include <stdlib.h>
3   #include <cstdlib>
4   #include <ctime>
5   #include <fstream>
6   #include <cmath>
7   #include "matrix.h"
8   #include "mpi.h"
9
10  using namespace std;
11  #define NUMBER_OF_RANDNUM 4000000
12  #define NUMBER_OF_SIMULATIONS 1000
13  #define NUMBER_OF_NODES 5
14
15  // --- Define constants for European option pricing ---
16  #define TIME_STEP 0.1
17  #define NUMBER_OF_STEPS 3650
18  #define EXPT_GW_RATE 0.0450 // Risk free interest rate
19  #define VOLATILITY 0.75
20  #define START_PRICE 1400
21  #define STRIKE_PRICE 1350
22  #define EXPIRE_DATE 365 // in days
23  #define START_DATE 0
24
25  int main(int argc, char* argv[]) {
26      char* rand_gauss_infile = "c:\\randgaus.txt";
27      char* option_price_file = "c:\\option_price.txt";
28
29      // --- Initialize MPI ---
30      int myrank, ranksize;
31      MPI_Status status;
32      MPI_Init(&argc, &argv);
33      MPI_Comm_rank(MPI_COMM_WORLD,&myrank);
34      MPI_Comm_size(MPI_COMM_WORLD,&ranksize);
35
36      // --- Split standard normal random variables ---
37      int duration = EXPIRE_DATE-START_DATE;
```

```
38      int Num_of_steps = int(double(duration)/double(TIME_STEP));
39      double R_V_gaussien;
40      double stockprices[int(NUMBER_OF_STEPS)+1];
41      double time_step; // in years
42      double options[NUMBER_OF_SIMULATIONS];
43      ifstream infile_gauss;
44      infile_gauss.open(rand_gauss_infile);
45
46      char node_rv_file[] = "c:\\rv_node0.txt";
47      int simu_per_node = NUMBER_OF_SIMULATIONS/NUMBER_OF_NODES;
48      double total = 0.0;
49
50      // --- if the node is the master ---
51      if (myrank == 0){
52        /* split normal random vector */
53        ifstream infile;
54        infile.open(rand_gauss_infile);
55        double rv;
56        for (int j=1; j<ranksize; j++){
57          ofstream outfile;
58          node_rv_file[10] = (char)(j+48);
59          outfile.open(node_rv_file);
60          for (int i=(j-1)*(NUMBER_OF_RANDNUM/NUMBER_OF_NODES)+1;
61                i<=j*(NUMBER_OF_RANDNUM)/NUMBER_OF_NODES;i++){
62            infile >> rv;
63            outfile << rv << " ";
64          }
65          outfile.close();
66        }
67        /* sent normal random vector */
68        for (int i=1; i < ranksize; i++){
69          node_rv_file[10] = (char)(i+48);
70          MPI_Send(&node_rv_file,15,MPI_CHAR,1,98,MPI_COMM_WORLD);
71        }
72        infile.close();
73      }
74      // --- if the node is a slave ---
75      else{
76        /* receive normal random vector */
77        MPI_Recv(&node_rv_file,15,MPI_CHAR,0,98,MPI_COMM_WORLD,&
                  ...status);
78
79        /* simulate Monte Carlo trajectories */
80        ifstream infile_gauss;
81        infile_gauss.open(node_rv_file);
82        for (int j=1; j<=simu_per_node; j++){
83          double time = 0;
84          stockprices[0] = double(START_PRICE);
85          time_step = double(TIME_STEP)/double(365);
86
87          int i=1;
88          while (i<=Num_of_steps && !infile_gauss.eof()){
89            infile_gauss >> R_V_gaussien;
90            stockprices[i]=stockprices[i-1]*
91              exp((EXPT_GW_RATE-pow(VOLATILITY,2)/2)*time_step
92                  +VOLATILITY*R_V_gaussien*sqrt(time_step));
```

```
93        time = time+time_step;
94        i = i+1;
95      }
96      /* write options price to array */
97      int steps;
98      steps = i;
99      double stockprice = stockprices[i-1];
100     if (stockprice >= double(STRIKE_PRICE))
101        options[j] = stockprice-double(STRIKE_PRICE);
102     else
103        options[j] = 0.0;
104    }
105    infile_gauss.close();
106    for (int i=1; i<simu_per_node; i++)
107       total = total+options[i];
108   }
109   // --- if the node is the master ---
110   if (myrank == 0){
111     /* collect results and compute */
112     double recvtotal = 0;
113     double recv = 0.0;;
114     double option_price_initial = 0.0;
115     for (int i=1; i<ranksize; i++){
116        MPI_Recv(&recv,2,MPI_DOUBLE,i,99,MPI_COMM_WORLD ,&status);
117        recvtotal = recvtotal+recv;
118     }
119     option_price_initial = exp(-1*EXPT_GW_RATE*double(duration)
            .../365)*(recvtotal/NUMBER_OF_SIMULATIONS);
120     ostream outfile;
121     outfile.open(option_price_file);
122     outfile << option_price_initial;
123     outfile.close();
124   }
125   // --- if the node is a slave ---
126   else{
127     /* Send result back to the master */
128     MPI_Send(&total,1,MPI_DOUBLE,0,99,MPI_COMM_WORLD);
129   }
130
131   MPI_Finalize();
132   return 0;
133  }
```

In lines 15–24, we define constants needed for European option pricing. The variable NUMBER_OF_RANDNUM contains the number of generated random numbers. The variable NUMBER_OF_SIMULATIONS contains the number of trajectories to be simulated per node. The variable NUMBER_OF_NODES contains the number of nodes in the grid. The variable TIME_STEP defines the discrete time step in days. The variable NUMBER_OF_STEPS contains the number of steps in a trajectory. The variable EXPT_GW_RATE contains the risk free interest rate. The variable VOLATILITY defines the volatility of the stock market. The variable START_PRICE defines the start price of the stock. The variable STRIKE_PRICE contains the strike price of the option. The variable EXPIRE_DATE defines the end date of the simulation. The variable START_DATE contains the starting date of the simulation.

In lines 29–35, we initialize the MPI.

In lines 36–14, if the node is a master node, we split the standard normal random variables into different files and send these file names to the slave nodes.

In lines 75–110, if the node is a slave node, we receive the normal random vector and simulate the Monte Carlo trajectories. The options price obtained at the end of each trajectory is written to an array. The total price is then computed by summing up the option prices.

In lines 111–128, if the node is a master node, we collect the results from the slave nodes and compute the final price. The final price is then written to a file.

In lines 129–135, if the node is a slave node, we send the total option price back to the master.

In the MPI/C++ *Globus* implementation, it is assumed that the standard normal random variables are already generated and stored in the file randgaus. txt. The master node then carries out the division of random variables among the nodes. Once this is done, the simulation of trajectories is made at each node and the total option computed at each node is sent back to the master node for processing. Lastly, the master node computes the overall option price at time 0 based on calculations from the individual nodes. The overall option price is then written to a file named option_price.txt.

7.5 Application to Nuclear Reactors in Computational Mechanics

In computational mechanics, the Monte Carlo method can be applied to many areas, such as the radiation of radioactive particles. Such radiation involves a highly random process. Due to the randomness of this phenomenon, this method can be used to simulate the effect of radiation on the resistivity of the material. In the following sections, we illustrate how the Monte Carlo method can be used to simulate such a phenomenon.

7.5.1 Nuclear Reactor-Related Criticality Calculations

In a nuclear reactor, the structure containing the reactor is often subjected to an intense amount of radiation due to nuclear fission reactions. These radioactive particles impact the surrounding mechanical structure and some areas may receive most of the radioactive impact. These areas increase the vulnerability of the structure, and thus these areas have to be reinforced in order to ensure the integrity of the surrounding structure. The simulation of the radioactive particle trajectories allows for the discovery of the points of weakness that could exist due to the deformation over time or a poor design of the structure.

7.5.2   Monte Carlo Methods for Nuclear Reactors

The radiation particles generated during the nuclear process can be considered to be random. Furthermore, the trajectories of the radioactive particles after they are produced are also highly random due to the presence of air particles that interact with these radioactive particle. Due to this phenomenon, Monte Carlo methods can be used to simulate the overall reaction and in particular the impact on the surrounding structure. Based on the properties of the surrounding environment, the radioactive particles emitted would follow a nondeterministic trajectory. Monte Carlo can be used to simulate different trajectories caused of the radioactivity in the nuclear reactor core.

7.5.3   Monte Carlo Methods for Grid Computing

The implementation in the grid architecture of Monte Carlo methods applied to nuclear reactors is quite similar to the example presented in option pricing. The inherent random property of the trajectories is provided by the random number generator discussed above in Section 7.3.1. Similarly, the total number of trajectories to be simulated is divided among various nodes to reduce the overall time needed. However, this example differs from the previous as it has a three dimensional environment, as compared to a one dimensional problem, which has an exercise time T as the boundary. Furthermore, we have also to consider the structural properties of the structure exposed to radiation. These mechanical properties have to be inculcated into the three dimensional boundary when implementing the Monte Carlo method. This is unlike the previous example where the boundary is just a vertical line. The definition of the structural properties for modeling in the grid is also beyond the scope of this book. However, we have seen here that, by running the Monte Carlo simulations, we will be able to have a distribution of the areas where the impact of the radioactive particles is the most intense. This allows us to identify potential weakness in the structure and help to prevent the compromise of the structure surrounding the nuclear reactor core.

In collaboration with the European union, several projects in the field of nuclear and particle physics are ongoing. They are namely the EGEE project, the UniGrids project, the Large Hadron Collider (LHC) computing grid project, and the LCG project. While the UniGrids project sees its use in various scientific areas such as computational biology, energy, geophysical depth imaging by oil companies, and reactor safety, the LHC project with CERN is highly useful in exploring new frontiers of science. This is possible thanks to the immense computational power offered by the grid to simulate scenarios that were previously impossible. The reader is referred to Chapter 6 for more details on these projects.

References

[149] Srinivas Aluru. Lagged Fibonacci random number generators for distributed memory parallel computers. *Journal of Parallel and Distributed Computing*, 45:1–12, 1997.

[150] Russel E. Caflisch and Suneal Chaudhary. Monte Carlo methods for american options. In *WSC'04: Proceedings of the 36th Conference on Winter Simulation*, pages 1656–1660, Washington, D.C., 2004. Winter Simulation Conference.

[151] John H. Halton. Sequential Monte Carlo techniques for solution of linear systems. *Journal of Scientific Computing*, 9(2):213–257, 1994.

[152] Nicole El Karoui. Couverture des risques dans les marchés financiers. Technical report, Ecole Polytechnique, CMAP, 91128 Palaiseau Cedex, France, 2003. Available online at: `http://www.cmap.polytechnique.fr/~elkaroui/index.html` (accessed January 1st, 2009).

[153] Spassimir H. Paskov. Computing high dimensional integrals with applications to finance. Technical Report CUCS-023-94, Columbia University, Department of Computer Science, New York, NY 10027, USA, 1994.

[154] Bernard Lapeyre and Emmanuel Temam. Competitive Monte Carlo methods for pricing of asian options. *Journal of Computational Finance*, 5(1):39–57, 2001.

[155] Fischer Black and Myron Scholes. The pricing of options and coporate liabilities. *The Journal of Political Economy*, 81(3):637–654, June 1973.

[156] Yongzeng Lai and Jerome Spanier. Applications of Monte Carlo/quasi-Monte Carlo methods in finance: option pricing. Technical report, Claremont Graduate University, Department of Mathematics, 925 N. Dartmouth Ave. Claremont, CA 91711, USA, 1999. Available online at: `http://citeseer.ist.psu.edu/264429.html` (accessed January 1st, 2009).

[157] John C. Hull. *Options, futures and other derivatives.* Prentice Hall finance series. Prentice Hall, Upper Saddle River, New Jersery 07458, 5th edition, 2002.

[158] Francis A. Longstaff and Eduardo S. Schwartz. Valuing american options by simulation: a simple least-squares approach. *The Review of Financial Studies*, 14(1):113–147, Spring 2001.

[159] Youcef Saad. *Iterative methods for sparse linear systems.* SIAM, 2003. 536 pages.

Chapter 8

Partial Differential Equations

8.1 Introduction

Partial differential equations (PDE) are a class of mathematical equation that is often encountered in many engineering problems. From Navier-Stokes equations in fluid mechanics to Black and Scholes equation in option pricing, PDE is a mathematical notion fundamental to various phenomena in the real world. Unfortunately, solving PDEs is often not straightforward and requires large computational resources that could be quite expensive. Computational resources, including computational time and memory, are often important assets that give a company the competitive edge over others. The advent of grid systems provides the industry with a new and creative solution to meet the ever-increasing demands for computational resources. Moreover, as PDEs play a central role in the modeling and simulation of various systems, solving PDEs with better speed and less memory is an interesting and attractive option for industry.

8.2 Deploying PDEs on Computational Grids

8.2.1 Data Parallelization

Data parallelization is the simplest method to be implemented in a grid system. To carry out such a parallelization, the domain of the parameters to the PDE is divided into many sub-domains. The data involved in each sub-domain are assumed to be independent. The validity of such a hypothesis is to be investigated for each problem otherwise the dependence of the parameters leads to false results.

For example, the modeling of fluid flows in a tube can be tested with various parameters of rugosity, viscosity, density, temperature, etc. The fluid flows are governed by the Navier-Stokes equations. Thus, to carry out these experiments, the same Navier-Stokes equations can be implemented on different nodes with different parameters (rugosity, viscosity, density, temperature, etc.) for each node of the grid. Since one node is independent of another node, by changing one parameter value in each different node, the whole experiment can be parallelized using the grid infrastructure. Hence, the total time of the overall experiment is reduced compared to running sequential experiments where only one parameter is changed at a time. This data parallelization method is essentially the same as running the overall experiment on a number of computers of which each has a different input parameter. This is the most basic way of using grids for industrial applications that require large numbers of parameters to be tested.

Figure 8.1 illustrates this approach for acoustic analysis within a car compartment for different frequencies.

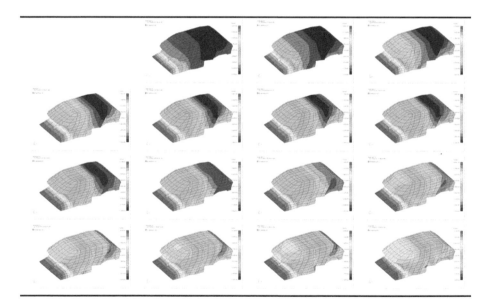

FIGURE 8.1: Acoustic simulation inside a car compartment. Acoustic pressure (in Pa) for different frequencies.

In the next two sections, we discuss two techniques, namely the parallelization in time and in space. However, these techniques can be done only after the PDE has been discretized both in time and space! In the following sections, we use the finite difference method in our discretization. Hence, we assume that discretization has already been carried out beforehand.

8.2.2 Time Parallelization

The parallelization method in time that we are going to employ in this section is called the parareal method; see reference [169, 170] for more details. The basic idea is to solve the system on two time scales with a different time step for each. The two time scales are namely the *rough* time scale and the *fine* time scale.

The overall time interval $]0, T[$ is first divided into N_r rough intervals, and in each interval the interval is further divided into N_f fine intervals. The system is first solved on the rough time scale. After this, for each rough interval, the system is solved again on the fine time scale. The solutions of the system on the fine time scale are independent and thus can be done in parallel. Once the fine solution on each interval is finished, the solution obtained through the rough solution is corrected and updated. The above procedure is repeated for k_{max} iterations where the final solution obtained converges towards the exact solution. We now describe how this method could be implemented in parallel.

Let $S(T_1, T_2, U_{T_1}, N)$ be the linear system obtained after the discretization of the problem where:

- T_1 denotes the initial time of the problem

- T_2 denotes the final time of the problem

- U_{T_1} denotes the discretization of the initial solution U_1

- N denotes the number of time steps used in this implicit method.

The complete algorithm could be written as follows:

- Step 1:

  The solution of $S(0, T, U(T_0), N_r)$ gives the discrete solution $U_r^0(T_i)$ at time T_i for $i = 1, \ldots, N_r$. This solution is the rough solution.

- Step 2:

  for $k = 0$ to k_{max}
  for $i = 0$ to N_{r-1}
  Solve $S(T_i, T_{i+1}, U_r^k(T_i), N_f)$
  end for
  The solution of the these linear systems are independent and thus can be done in parallel. With these solutions, the discrete solution $U_f^k(T_i)$ for $i = 1, \ldots, N_c$ can be found.

  The correction of the rough solution is obtained by the following equation: $J^k(T_i) = J^{k-1}(T_i) + U_f^k(T_i) - U_r^k(T_i)$ with $J^{-1}(T_i) = 0$ for

$i = 1, \ldots, N_r$.

With these corrections, the quantity $S(0, T, U(T_0), N_r)$ is re-evaluated by applying the corrections at each rough time step.

The correction is made by adding $J^k(T_i)$ to the corresponding rough solution $Uk_r(T_i)$ at each time step.

    end for

- Step 3:

  $U_r^{k_{max}}(T)$ is the approximative solution of the system. It is the solution $U(T)$ when $k = k_{max}$.

Thus, we have seen with the parareal method, how the solution of the PDE can be found in parallel. This method can be easily implemented on the parallel structure of a grid. The mathematical proof of the convergence of this method and the error refinement are discussed in details in reference [169, 170].

8.2.3   Spatial Parallelization

In this section, we describe how the computational domain of the PDE can be subdivided into smaller domains for parallel computations.

Mathematical Formulation

For the sake of simplicity we present the principles of a two dimensional heat problem. This problem could be expressed as:

$$\frac{\partial u}{\partial t}(t, x, y) - \nu \Delta u(t, x, y) = g(t, x, y) \ in \ ]0, T[\times \Omega, \tag{8.1}$$

$$u(t = 0, x, y) = 0 \ in \ \Omega, \tag{8.2}$$

$$u(t, x, y) = 0 \ on \ [0, T] \times \partial \Omega \tag{8.3}$$

where $\Omega =]-1, 1[^2$ and $T = 1$. The quantity Δu denotes the Laplacian defined by $\Delta u = \frac{\partial^2 u}{\partial x^2} + \frac{\partial^2 u}{\partial y^2}$. The thermal conductivity coefficient is set to $\nu = 0.01$ and the mobile heat source term is given by the relation:

$$g(t, x, y) = \chi(\| (x, y)^T - (\frac{1}{2} \cos 4\pi t, \frac{1}{2} \sin 4\pi t)^T \| < \frac{1}{4}) \tag{8.4}$$

where χ is the characteristic function defined by:

$$\chi(\| z \| < \frac{1}{4}) = 1 \quad if \quad \| z \| < \frac{1}{4}$$

$$\chi(\| z \| < \frac{1}{4}) = 0 \quad otherwise.$$

Here the source of the heat problem is moving in a circle around the center of the domain. The evolution of the domain's temperature upon the time is determined through the solution of the PDE equation 8.1.

Sequential Solution of the Heat Equation

As mentioned earlier, the PDE must be discretized. For this purpose a regular mesh of mesh size h is considered for the spatial discretization, as represented in Figure 8.2. The coordinates $(x_i, y_i) = x_{i,j}$ of the points of the

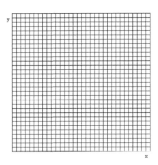

FIGURE 8.2: Two dimensional spatial discretization.

mesh are given by:

$$x_{i,j} - (-1 + ih, -1 + jh)^T, \quad \text{for} \quad i, j = 1, \ldots, M \tag{8.5}$$

with $M = 100$ and $h = \frac{2}{M+1}$. A constant time step equal to Δt such as $t^{n+1} = t^n + \Delta t$ is considered for the time discretization. The quantity u_{ij}^n denotes the numerical approximation of the continuous solution u at the coordinate point $(x_i, y_i) = x_{i,j}$ at the time t^n.

With these notations, the explicit numerical schema considered is the following:

$$u_{ij}^{n+1} = u_{ij}^n + \frac{\nu \Delta t}{h^2}(u_{i-1,j}^n + u_{i+1,j}^n + u_{i,j-1}^n + u_{i,j+1}^n - 4u_{ij}^n) + \Delta t g_{ij}^n \tag{8.6}$$

where $g_{ij}^n = g(t^n, x_{ij})$. This means that the discrete solution u at the time t^{n+1} can thus be computed using the previous computed approximate solution at the time t^n. Figure 8.3 represents the solution of the heat equation at different time steps. A zoom of the final solution at time $t = T$ is shown in Figure 8.4.

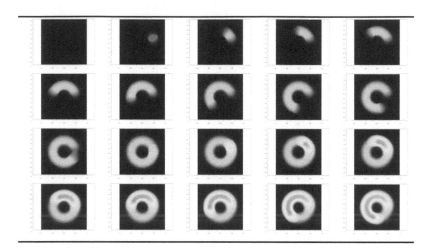

FIGURE 8.3: Evolution of the domain upon the time.

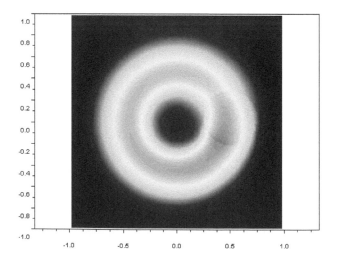

FIGURE 8.4: Solution of the heat equation at time T.

Parallel Solution of the Heat Equation

The spatial parallelization can be carried out by dividing the two-dimensional mesh into smaller sub-meshes as represented in the Figure 8.5 in the case of four sub-meshes. The solution of the heat equation on each

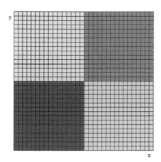

FIGURE 8.5: The discretized domain is divided into four sub-domains and distributed among four nodes for parallel computation.

sub-mesh could be performed on different nodes of the grid.

At each time step, the results computed from each node will be gathered to produce the overall results of u at time t^{n+1}. This solution is then used as the input parameter for computing the solution at time t^{n+2} at each node. Repeating this procedure, the solution of u_{ij}^{n+1} for $t = 0, \Delta t, 2\Delta t, \ldots, T$ can be obtained.

As we can see in Figure 8.6, the solution at time step t^n in the global domain is divided into four sub-domains. It is important to note that during this division process, the sub-domains overlap each other. This is essential for the computation of the solution at the next time step t^{n+1}. Indeed, the overlapped area of each sub-domain provides the boundary condition of the neighboring sub-domains. By fulfilling the boundary condition of each sub-domain, the continuity of the solution is ensured. Once the dividing process is finished, the divided sub-domains are passed to the respective nodes for processing the sub-solution at the next time step t^{n+1}. Since the condition of continuity is ensured, the sub-solutions computed by each node can simply be pieced together to produce the overall solution. On the left of Figure 8.6, the solution at time t^n is divided into four equal sub-domains numbered 1,2,3,4. These are the sub-domains of the solution that are going to be computed by the respective nodes 1,2,3,4. As mentioned earlier, the data passed to each node is bigger than each sub-domain due to the overlapping area. Once the required data is sent to each of the respective nodes, computation of the sub-solution at the next time step t^{n+1} is carried out. The explicit schema (8.6)

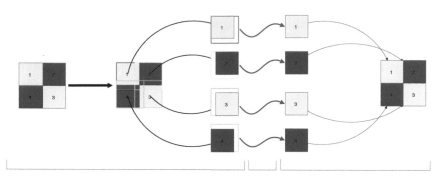

FIGURE 8.6: Parallelization in space.

is applied here. This is shown in the center of the figure. The curved arrows denote the computation of the sub-solution at the next time step t^{n+1}. Since this computation considers the boundary condition of the sub-domains, the continuity between the sub-solutions at the next time step t^{n+1} is ensured. Lastly, on the right part of Figure 8.6, the sub-solutions are pieced together to produce the final solution at the next time step t^{n+1}. It is important to note that for the explicit method to be stable, the time step Δt is here chosen to be $0.95(\frac{h^2}{4\nu})$.

REMARK 8.1   In the program code provided later in this chapter, the message passing of one divided sub-part to the neighboring sub-domains is not carried out. Instead (and only for the sake of simplicity of the presented program) the overall solution at time step t^n is passed to each of the neighboring sub-domains.   ⬜

In this section, we have briefly seen the spatial parallelization of the explicit finite difference method used on a two dimensional heat equation. While the illustration here is limited to a two dimensional problem, such a parallelization procedure can be generalized to higher dimensional problems without much difficulty.

8.3   Application to Options Pricing in Computational Finance

In this section, we see how the discretization and solution process of PDEs can be applied in parallel to the area of computational finance, notably in the pricing of options. For this purpose, the Black and Scholes equation, which governs the evolution of the price of options over time, is considered.

8.3.1 Black and Scholes Equation

We discuss here the valuation of European options with many underlying stocks. The owner of such an option has the right to buy a combination of stocks specified in the contract at the strike price K and at the exercise time T. The number of underlying stocks d gives the dimension of the problem. In our case, we go through a preprocessing procedure to transform the Black and Scholes equation into the heat equation, as initially proposed in reference [161]. Many numerical methods exist for solving the heat equation. Upon these methods, radial basis functions could be used to approximate the solution due to the symmetry of the heat equation. However, this approximation exploiting symmetry is not presented here as it is beyond the scope of the book.

Some notations are now introduced. The quantity $F(t, \mathbf{s})$ denotes the value of the option at time t with the underlying stock prices $\mathbf{s} = (s_1, \ldots, s_d)^T$. It is also the solution to the following boundary value problem:

$$\frac{\partial F}{\partial t} + r \sum_{i=1}^{d} s_i \frac{\partial F}{\partial s_i} + \frac{1}{2} \sum_{i,j=1}^{d} [\sigma\sigma^T]_{ij} s_i s_j \frac{\partial^2 F}{\partial s_i \partial s_i} \quad rF = 0, \quad \forall t < T, \ \mathbf{s} \subset \mathbb{R}_+^d,$$

$$(8.7)$$

$$F(T, \mathbf{s}) = \Phi(\mathbf{s}), \quad \forall s, \mathbf{s} \in \mathbb{R}_+^d, \tag{8.8}$$

$$F(t, \mathbf{s}) \to \frac{1}{d} \sum_{i=1}^{d} s_i - Ke^{-r(T-t)}, \quad \| \mathbf{s} \| \to \infty \quad \forall t \tag{8.9}$$

$$F(t, \mathbf{s}) \to 0, \| \mathbf{s} \| \to 0, \quad \forall t \tag{8.10}$$

where r is the risk-free interest rate and σ the volatility matrix; see [163]. The contract function $\Phi(\mathbf{s})$ is defined as

$$\Phi(\mathbf{s}) = \max(0, \frac{1}{d} \sum_{i=1}^{d} s_i - K). \tag{8.11}$$

We rewrite equations (8.7)-(8.8) as an initial value problem upon the new variable $\tau = T - t$. This gives us the following equation.

$$\frac{\partial F}{\partial \tau} = \frac{1}{2} \sum_{k=1}^{d} \sigma_k^T S \nabla_s (\nabla_s F)^T S^T \sigma_k + rs^T \nabla_s F - rF, \quad \tau > 0, \ \mathbf{s} \in \mathbb{R}_+^d, \tag{8.12}$$

$$F(0, \mathbf{s}) = \Phi(\mathbf{s}), \quad \mathbf{s} \in \mathbb{R}_+^d, \tag{8.13}$$

where σ_k is the k-th column of the volatility matrix and where

$$S = S^T = diag(\mathbf{s}).$$

Similarly, rewriting equations (8.9)-(8.10) upon the variable τ gives

$$F(\tau, \mathbf{s}) \to \frac{1}{d} \sum_{i=1}^{d} s_i - K e^{-r\tau}, \quad \| \mathbf{s} \| \to \infty \tag{8.14}$$

$$F(\tau, \mathbf{s}) \to 0, \| \mathbf{s} \| \to 0. \tag{8.15}$$

Next, we carry out the following transformation:

$$s = e^{(A^T(Q^T x + b))}. \tag{8.16}$$

With the above transformation and under the constraint

$$A^T A = \frac{1}{2} \sum_{k=1}^{d} (\sigma_k \sigma_k^T)$$

Equation (8.12) reduces to:

$$\frac{\partial F}{\partial \tau} = \Delta_x F + \alpha^T (QA)^{-1} \nabla_x F - rF. \tag{8.17}$$

The final transformation is to let

$$F(\tau, \mathbf{x}) = e^{\gamma \tau + \xi^T x} P(\tau, \mathbf{x})$$

where $\xi^T = -\frac{1}{2} \alpha^T (QA)^{-1}$ and $\gamma = \alpha^T (QA)^{-1} \xi + \xi^T \xi - r = -\xi^T \xi - r$. In summary, the multidimensional option pricing partial differential equation has been reduced to the following system of equations:

$$\frac{\partial P}{\partial \tau} = \Delta_x P, \quad \mathbf{x} \in \Omega, \quad \tau > 0,$$

$$P(\tau, \mathbf{x}) = e^{-\gamma \tau - \xi^T x} F(\tau, \mathbf{x}) \equiv f(\tau, \mathbf{x}), \quad \mathbf{x} \in \Gamma_D, \quad \tau > 0, \tag{8.18}$$

$$P(0, \mathbf{x}) = e^{-\gamma \tau - \xi^T x} \Phi(\mathbf{x}) \equiv f_0(\mathbf{x}) \ \mathbf{x} \in \Omega,$$

where Ω is the computational domain and Γ_D is the boundary of this transformed computational domain Ω. The reader familiar with functional analysis will recognize the heat equation. The computational domain Ω can be tailored to make it symmetrical. First, the computational domain in the original s-coordinates has to be truncated to make it finite for practical implementations. For this, we choose

$$R = s \mid \frac{1}{d} \sum_{i=1}^{d} s_i \le 4K, s_i \ge 0, i = 1, \dots, d.$$

From this computational domain in s-coordinates, we apply the transformation (8.16), rotation Q and translation b in such a way that the region R is mapped onto a hypercube $[-L, L]^d$. With this consideration, the computational domain in the x-coordinate is symmetrical. In two spatial dimensions, the hypercube conforms to a square as in the example shown below.

8.3.2 Discrete Problem

The discretization of the above heat equation can be carried out both in time and space as described in Sections 8.2.3-8.2.2. The solution of the discrete system of equations can be performed in parallel using a grid infrastructure. For the numerical experiments, we take $\nu := 1$, $g(t, \cdot) := 0$, and the computational domain is of the form

$$R = \{s | \frac{1}{d} \sum_{i=1}^{d} s_i \leq 4K, s_i \geq 0, i = 1, \ldots, d\}$$

where d is the dimension of the domain.

8.3.3 Parallel Solution of Black and Scholes Equation

The parallel solution of the Black and Scholes equation can be done by solving a heat equation in parallel. The discretized domain is split into several nodes. Each node will then be responsible for computing the solution for a section of the domain. However, as mentioned in Section 8.2.3, the part of the solution passed to each node has to include the overlapped area as illustrated in Figure 8.6 to ensure the continuity of the overall solution at the next time step t_{n+1}.

As indicated previously, in Figure 8.7, the whole solution instead of the sub-solution at time t_n is passed as a parameter to each node in the grid. There is no loss of generality as the whole solution effectively includes the sub-solution mentioned in Section 8.2.3. From the bottom of the figure, the solution at time t_n is passed as a parameter to each of the nodes. Each node is then responsible for computing a sub-solution at time t_{n+1} based on the solution at time t_n. The sub-solutions are then pieced together to form the overall solution at time t_n.

We illustrate such a methodology on a C++ program using the Message Passing Interface (MPI) library. It is important to notice that the main motivation of using MPI is to facilitate the eventual migration of the C++ program to the *Globus* infrastructure. Indeed, such an implementation can be migrated to *Globus* without any changes as *Globus* uses the same MPI structure to run parallel computation. The complete program is illustrated in Listing 8.1.

Listing 8.1: Parallel solution of the heat equation.

```
1   #include "mpi.h"
2   #include <cmath>
3
4   // --- Define constants for the heat equation ---
5   #define M 10
6   #define T 1
7   #define NU 0.01
```

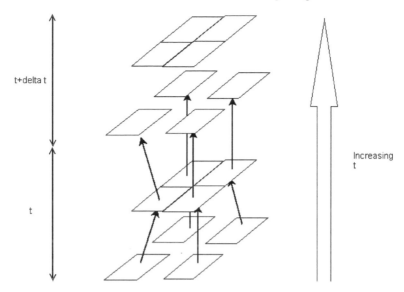

FIGURE 8.7: The discretized domain is computed using four parallel nodes and the data is gathered at the master nodes. This process is then repeated for subsequent time steps.

```
 8   #define BOUNDARY1 -1
 9   #define BOUNDARY2 1
10   #define PI 3.14159265
11
12   // --- Define number of nodes of the grid ---
13   #define NUM_OF_NODESX 2
14   #define NUM_OF_NODESY 2
15
16   // --- Define mobile source term ---
17   double g (int i, int j, double h, double time){
18      double x = (-1)+j*h;
19      double y = (-1)+i*h;
20      double dist=0;
21      return (4*sqrt(
22          (x-(0.5*cos(4*PI*time)))*(x-(0.5*cos(4*PI*time)))
23        + (y-(0.5*sin(4*PI*time)))*(y-(0.5*sin(4*PI*time)))) < 1);
24   }
25
26   // --- Numerical schema ---
27   double schema(double Unow[M+2][M+2], int i, int j, double h,
           ...double time, double deltat){
28      return Unow[i][j] + deltat*g(i,j,h,time)
29        + ((double(NU)*deltat)/(h*h))
30        * (Unow[i-1][j] + Unow[i+1][j] + Unow[i][j-1] + Unow[i][j
           ...+1]-4*Unow[i][j]);
31   }
32
```

```
33   int main(int argc, char* argv[]) {
34     // --- Initialization of MPI ---
35     int myrank;
36     int ranksize;
37     MPI_Status status;
38     MPI_Init(&argc,&argv);
39     MPI_Comm_rank(MPI_COMM_WORLD,&myrank);
40     MPI_Comm_size(MPI_COMM_WORLD,&ranksize);
41
42     // --- Initialization of program parameters ---
43     double Unext[M+2][M+2];
44     double Unow[M+2][M+2];
45
46     for (int i=0; i<M+2; i++)
47       for (int j=0; j<M+2; j++){
48         Unext[i][j] = 0;
49         Unow[i][j] = 0;
50       }
51     double h=(double(BOUNDARY2)-double(BOUNDARY1))/(double(M)+1);
52     double deltat = 0.95*((pow(h,2))/(4*NU));
53
54     // --- Computation of solution iteratively ---
55     double time = 0;
56     int msgid = 98;
57     while (time <= T){
58       /*Sending out data to slave nodes for processing
59         Sending data back to master node by the slaves*/
60
61       if (myrank == 0){
62         for (int i=1; i<ranksize;i++)
63           MPI_Send(&Unow,(M+2)*(M+2),MPI_DOUBLE,i,msgid,
                    ...MPI_COMM_WORLD);
64       }
65       else{
66         MPI_Recv(&Unow,(M+2)*(M+2),MPI_DOUBLE,0,msgid,
                  ...MPI_COMM_WORLD,&status);
67         if (myrank == 1){
68           for (int i=1; i<=int(M)/int(NUM_OF_NODESX); i++)
69             for (int j=1; j<=int(M)/int(NUM_OF_NODESY); j++)
70               Unext[i][j] = schema(Unow,i,j,h,time,deltat);
71         }
72         else if (myrank == 2){
73           for (int i=int(M)/int(NUM_OF_NODESX)+1; i<=int(M); i++)
74             for (int j=1; j<=int(M)/int(NUM_OF_NODESY); j++)
75               Unext[i][j] = schema(Unow,i,j,h,time,deltat);
76         }
77         else if (myrank == 3){
78           for (int i=1; i<=int(M)/int(NUM_OF_NODESX); i++)
79             for (int j=int(M)/int(NUM_OF_NODESY)+1;j<=int(M);j++)
80               Unext[i][j] = schema(Unow,i,j,h,time,deltat);
81         }
82         else if (myrank == 4){
83           for (int i=int(M)/int(NUM_OF_NODESX)+1; i<=int(M); i++)
84             for (int j=int(M)/int(NUM_OF_NODESY)+1;j<=int(M);j++)
85               Unext[i][j] = schema(Unow,i,j,h,time,deltat);
86         }
```

```
 87          }
 88
 89          MPI_Barrier(MPI_COMM_WORLD);
 90
 91          msgid = msgid +1;
 92          if (myrank == 0){
 93             for (int i=0; i<M+2; i++)
 94                for (int j=0; j<M+2; j++)
 95                   Unow[i][j] = 0;
 96             for (int k=1; k<ranksize; k++){
 97                MPI_Recv(&Unext,(M+2)*(M+2),MPI_DOUBLE,k,msgid,
                       ...MPI_COMM_WORLD,&status);
 98                for (int i=1; i<=int(M); i++)
 99                   for (int j=1; j<=int(M); j++)
100                      Unow[i][j] = Unow[i][j]+Unext[i][j];
101             }
102          }
103          else{
104             MPI_Send(&Unext,(M+2)*(M+2),MPI_DOUBLE,0,msgid,
                       ...MPI_COMM_WORLD);
105          }
106          time=time+deltat;
107          msgid=msgid+1;
108       }
109
110       MPI_Finalize();
111       return 0;
112    }
```

In lines 4–11, we define constants for the heat equation. The variable M defines the number of discrete points in each direction. The variable T defines the time interval of the simulation. The variable NU defines the heat transfer coefficient. The variable BOUNDARY1 defines one boundary of the computational domain. The variable BOUNDARY2 defines the other boundary in the computational domain. The computational domain is symmetrical in the x and y directions. The variable PI defines the value of π.

In lines 12–15, we define the number of nodes in the x and y direction as NUM_OF_NODESX and NUM_OF_NODESY, respectively.

In lines 16–25, we define the function g() as the mobile source term.

In lines 26–35, we define the function schema() for the explicit finite difference method.

In lines 34–41, we initialize the MPI.

In lines 42–53, we initialize the program parameters. The variable Unext is the solution at the next time step. The variable Unow is the solution at the present time step. The variable h is the width between each discrete point in the x ad y axis. The variable deltat is the width between each time step.

In lines 54–110, we compute the solution at each time step using a while loop that iterates from $time = 0$ to $time = T$. At each time step, the master node sends out the solution of the previous time step to the slave nodes for processing. Each slave node computes a sub-solution for the next time step. The sub-solution computed by each slave node is sent back to the master node. The master node then assembles the sub-solutions from the slave nodes to produce the global solution.

In lines 58–64, the master node sends out the solution at the present time step to the nodes for processing.

In lines 65–88, the slave nodes compute a section of the solution for the next time step based on the identity of the node.

In lines 92–102, the master node receives the data from the slave nodes and the data is combined to obtain the solution at a particular time step.

In lines 103–106, the slave node sends out the computed sub-solutions at the next time step to the master node.

This is a parallel implementation on MPI/$C++$ using four computational nodes. In the cluster, there is also a master node that helps to gather and broadcast information to the other nodes.

During the computation of the solution, the explicit method is used in the finite difference method. This means that the solution U at time t_{n+1} is based solely on the solution U at time t_n. Hence, using parallel computation on four nodes, the solution U at time t_n is sent to the individual nodes. Furthermore, the computational domain can be divided into four sub-domains. In this way, each node is responsible for computing a quarter of the solution at the next time step. This means that each of these sub-domains will then be computed by each node to obtain the sub-solution of U at time t_{n+1}. These four sub-solutions will then be sent back to the master node where the sub-solutions will be combined to obtain the solution U at time t_{n+1}. This process is repeated until the time boundary is reached.

References

[160] Ioane Muni Toke. *Résolution de modèles d'évaluation de produits dérivés financiers sur des architectures de grilles informatiques.* PhD thesis, Ecole Centrale Paris, MAS, 92295 Chatenay Malabry Cedex, France, September 2006.

[161] Elisabeth Larsson, Krister Ahlander, and Andreas Hall. Multi-dimensional option pricing using radial basis functions and the generalized Fourier transform. Technical Report 2006-037, Uppsala University, Department of Information Technology, August 2006.

[162] John C. Hull. *Options, futures and other derivatives.* Prentice Hall finance series. Prentice Hall, Upper Saddle River, New Jersery 07458, 5th edition, 2002.

[163] Damien Lamberton and Bernard Lapeyre. *Introduction to stochastic calculus applied to finance.* Chapman & Hall, CRC Press, 1st edition, 1996.

[164] Fischer Black and Myron Scholes. The pricing of options and coporate liabilities. *The journal of political economy*, 81(3):637–654, June 1973.

[165] Angelika Esser. Derivatives written on a power of stock price: general valuation principles and application to stochastic volatility models. Technical report, Goethe University, School of Business and Economics, Mertonstr. 17, D-60054 Frankfurt am Main, Germany, 2003. Available online at: http://ssrn.com/abstract=356007 (accessed January 1st, 2009).

[166] Ulrika Pettersson, Elisabeth Larsson, Gunnar Marcusson, and Jonas Persson. Improved radial basis functions methods for multi-dimensional option pricing. Technical Report 2006-028, Uppsala University, Department of Information Technology, May 2006.

[167] Francisco Chinesta, Antonio Falco, and Mariano Gonzalez. Model reduction methods in option pricing. Technical Report WP-AD 2006-16, CNRS, LMSP, UMR 8016, 151 Boulevard de L'Hôpital, 75013 Paris, France, 2006. Available online at: http://www.ivie.es/downloads/docs/wpasad/wpasad-2006-16.pdf (accessed June 12th, 2007).

[168] Yiu chung Hon and Xian zhong Mao. A radial basis function method for solving options pricing model. *The Journal of Financial Engineering*, 8(1):31–49, 1999.

[169] Yvon Maday and Gabriel Turinici. A parareal in time procedure for the control of partial differential equation. *C. R. Acad. Sci., Série I, Math.*, 335(4):387–392, 2002.

[170] Jacques-Louis Lions, Yvon Maday, and Gabriel Turinici. Résolution d'edp par un schéma en temps "pararéel". *C. R. Acad. Sci., Série I, Math.*, 332(7):661–668, 2001.

Appendix A

Globus

A.1 Overview of Globus Toolkit 4

The open source *Globus Toolkit (GT)* provided by Globus Alliance [171] is a fundamental tool for building grids and realizing grid applications. The current released version is GT4. GT includes resource monitoring and discovery services, job submission infrastructure, security infrastructure and data management services. GT implements the following standards:

- *Open Grid Services Architecture (OGSA)*

- *Open Grid Services Infrastructure (OGSI)*

- *Web Services Resource Framework (WSRF)*

- *Job Submission Description Language (JSDL)*

- *Distributed Resource Management Application API (DRMAA)*

- *WS-Management*

- *WS-BaseNotification*

- *SOAP*

- *WSDL*

- *Grid Security Infrastructure (GSI)*

GT provides a set of components implemented on top of WSRF. It supports the development of new web services using Java, C and Python. Besides the WSRF-based components, GT also includes components that have not been implemented on top of WSRF (such as GridFTP). These components are called *pre-WS services*. GT offered two ways, client APIs (in different languages) and commands, to access these services. These components realize the functions of security, data management, execution management, information services, and common runtime.

A.2   Installation of Globus

The latest stable release of GT4 available at the time of writing this book is version 4.0.4. You can download GT4 from `http://www.globus.org/toolkit/downloads/`. While downloading you can choose from a list of various releases and platforms on which you want to install the Globus Toolkit. For the purpose of this discussion we assume that you have downloaded version 4.0.0 of Globus Toolkit on your machine. The GT4 installation manual lists some software prerequisites for the GT4 installation. These are:

- J2SE 1.4.2+

- Ant 1.5.1+ (or 1.6.1+ if you are using Java 1.5)

- A C compiler, preferably `gcc`

- Sone utilities: `tar`, `make` and `sed`

- `zlib`: A compression/decompression tool available at `http://www.zlib.net/`

- A JDBC compliant database

After installation of this software, you can proceed to the installation of GT4.

1. We create a non-privileged user named `globus` for performing administrative tasks such as starting and stopping the globus container. Create an installation directory and assign 'read' and 'write' privileges on that directory to the user `globus`.

2. Apache Ant uses the Java referred to by the environment variable `JAVA_HOME`. This variable should be set to the top-level directory of your Java installation. If your operating system already comes bundled with an Ant software, it might be preconfigured to use the `gcj` compiler. Change that to the J2SE java compiler. The configuration file for Ant is `/etc/ant.conf`.

3. For the purpose of this installation guide we assume that Globus is installed at `/usr/local/globus-4.0.0`. You can, however, change the installation directory to suit your needs. Run the following commands as `globus` user:

```
export GLOBUS_LOCATION=/usr/local/globus-4.0.0
./configure --prefix=$GLOBUS_LOCATION
```

./configure provides a list of custom install options. The complete list of options can be seen using ./configure--help. It provides options for building the GRAM interface for various schedulers, such as Condor, PBS, LSF, and provides option to install certain services.

4. Run the make command as a Globus user. This can take a lot of time depending on the configuration of your machine and the packages you choose to install.

5. Finally run the command makeinstall to install GT4. This completes the installation of Globus Toolkit 4 on your machine.

After installing the GT4, you need to configure the security parameters to get started. If your organization already has a certifying authority (CA), you can request your X.509 certificate from it. If not, the Globus Toolkit comes bundled with a SimpleCA package. You can configure SimpleCA to issue digital certificates to your user. Configuring SimpleCA and requesting certificates from it has been described in Section 4.3.1 entitled "Getting a Certificate". For details of configuring other services such as GridFTP, GSI-OpenSSH, MyProxy etc., you can refer to the GT4 admin guide available online at http://www.globus.org/toolkit/docs/4.0/admin/docbook/admin.pdf.

A.3 GT4 Configuration

GridFTP Configuration

When you install GT4, GridFTP is installed by default. It can function without specific configuration. Run the following command to start GridFTP server, with -S to make the command run as a daemon, and -p2811 to specify the listening port for client connections:

```
$GLOBUS_LOCATION/sbin/globus-gridftp-server -S -p 2811
```

RFT Configuration

RFT is used to perform third-party transfers across GridFTP servers. GridFTP records the transfers status through RTF so that they can be recovered from failures. As these records are stored in a Database, a Database should be installed on the system. In this annex, we use PostgreSQL v8.1.4 as the Database. The installation and configuration of PostgreSQL documents are available at http://www.postgresql.org/docs/manuals/

1. Run the following command as root, to configure the postmaster daemon, accept to TCP connections, by adding -o -i options.

```
postmaster -o -i -D PostgreSQL_INSTALL_PATH/data/
```

2. Run `createdbrftDatabase` to create the database used for RFT. Then, create the database tables with the sql script offered by GT4:

   ```
   psql -d rftDatabase -f $GLOBUS_LOCATION/share/
   globus_wsrf_rft/rft_schema.sql
   ```

3. Modify file $GLOBUS_LOCATION/etc/globus_wsrf_rft/jndi-config.xml as globus. Find the item `connectionString` in the<resource> section and change the values of `driverName, connectionString, userName, password`.

GRAM Configuration

WS GRAM is installed as part of a default GT 4.0 installation.

WS GRAM executes and manages jobs through the local scheduler. GT4 supports the local schedulers PBS, Condor, LSF. One of these local schedulers should be installed and configured in advance. RFT must be installed and configured because the file-staging of WS GRAM depends on RFT.

Configuration

1. `sudo` configuration: As `root`, open file `/etc/sudoers` and add the following lines:

   ```
   globus ALL=(username1,username2) NOPASSWD:
       /usr/local/globus-4.0.1/libexec/globus-gridmap-and-
       ...execute
       /opt/globus/GT3.9.4/libexec/globus-job-manager-script.pl
       ... *
   globus  ALL=(username1,username2) NOPASSWD:
       /opt/globus/usr/local/globus-4.0.0/libexec/globus-
       ...gridmap-and-execute
       /opt/globus/usr/local/globus-4.0.0/libexec/globus-gram-
       ...local-proxy-tool *
   ```

   In these lines, `/usr/local/globus-4.0.0/` should be replaced with the install path $GLOBUS_LOCATION.

2. Make the host credentials accessible by the container

   ```
   % cd /etc/grid-security
   % cp hostcert.pem containercert.pem
   % cp hostkey.pem containerkey.pem
   % chown globus.globus container*.pem
   ```

3. Local scheduler adapter configuration (example of PBS)

   Install PBS adapter:

   ```
   % cd $GLOBUS_LOCATION\gt4.0.0-all-source-installer
   % make gt4-gram-pbs
   % make install
   ```

Configure the remote shell for rsh:

```
% cd $GLOBUS_LOCATION/setup/globus
% ./setup-globus-job-manager-pbs --remote-shell=rsh
```

A.4   Main Components and Programming Model

A.4.1   Security (GSI)

Grid security is used to establish the authentication for users and services. It provides secure communication, authorization of permitted actions for grid users, management of users' credentials and membership information. GT4's security mechanism is based on X.509 end entity certificate and proxy certificate.

Grid Security Infrastructure (GSI) is a security component of GT. It provides authentication and message protection for the communication between users and grid resources or services. The major motivations of GSI are:

- Establishing secure communication (authenticated and confidential) between grid elements;

- Supporting security across organizational boundaries without using a centralized security system;

- Supporting single sign-on for grid users (to avoid re-entering passphrase when user's application uses multiple grid sites);

GSI has the following functions.

Mutual Authentication

Mutual authentication means the process that two entities verify the identity of each other and recognize each other's certificates. Mutual authentication is based on the Secure Socket Layer (SSL) protocol. The precondition of mutual authentication is that both the entities trust each other's CA[1]. So both entities have a copy of the certificate of the other entity's CA. They must establish correctly that these certificates belong to the respective CAs. This is a symmetrical process. Figure A.1 illustrates this process.

The first phrase is the authentication from B to A. A sends its certificate to B. B confirms A's certificate and sends a test message to A. A receives this message ,encrypts this message with A's own private key and sends it to B. In order to test A's identity, B decrypts this message with A's public key stored

---

[1]CA is used to sign the certificate of the other entity; see Chapter 4.

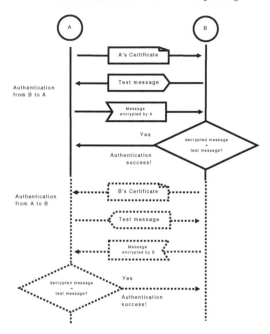

FIGURE A.1: Mutual authentication.

in A's certificate. If the decrypted message matches with the original message, B trusts A. The second phrase involves the same authentication process, but this time from A to B. Through mutual authentication both the parties, A and B, trust each other.

Confidential Communication

Confidential communication is an optional security feature. It is done by establishing a secret key for encryption. With this key, the two parties can send and receive encrypted messages. A related security feature is communication integrity, which does not require encryption of the messages. Communication integrity mechanism means that even the eavesdropper can read the private message between two parties but cannot modify the message. Communication integrity causes some extra overhead. However, this overhead is smaller than the overhead in encryption.

Securing Private Keys

A private key is important for authentication and encryption of messages. The user's private key is stored in a file in the local computer's filesystem. In order to protect this file from unrestricted access, it is encrypted with a password. To use the key, the user must enter the password required to decrypt the file containing the private key.

Delegation, Single Sign-On and Proxy Certificates

When the user's application involves several different resources user has to re-enter the password many times. GSI provides a proxy credential for avoiding this trouble. Similar to a user credential, a proxy credential also contains a certificate and a private key. However it is slightly different from the user's credentials so that an application knows that this proxy is on the user's behalf and not the actual user. The proxy certificate is signed by its owner, rather than a CA. Once a proxy is created and stored, the user can use the proxy certificate and private key for mutual authentication without entering a password. The format of a proxy certificate is discussed in Chapter 4.

A.4.2 Data Management (RFT)

GT4 contains several components for data management. GridFTP and *Reliable File Transfer (RFT)* service are used for data movement. *Replica Location Service (RLS)* takes charge of tracing the copies or replicas of data. *Data Replication Service (DRS)*, based on RFT and RLS, is responsible for a pull-based replication capability for grid files. Among these components, GridFTP and RLS are the pre-WS components. RFT and DRS belong to the WS components group.

Data Movement Components: GridFTP and Reliable File Transfer (RFT) Service

GridFTP is a pre-WS component. It consists of three parts:

- `globus-gridftp-server`, installed in a server host where the data resource exists.

- `globus-url-copy`, a scriptable command line client for globus-gridftp-server;

- A set of libraries.

Using the `globus-gridftp-server`'s interface, different file systems can be accessed by the same client tool, `globus-url-copy`. This interface is called Data Storage Interface. POSIXfile system, *Storage Resource System (SRB)*, High Performance Storage System (HPSS) and NeST (used in Condor) are supported by this interface. `globus-url-copy` can access the data resource by a number of protocols including http, https, ftp, gsiftp, and file. An example is shown below:

```
globus-url-copy gsiftp://remote.host.edu/path/to/file
file:///path/on/local/host}
```

This command transfers a remote file to the indicated directory of a local host.

GridFTP is an effective tool for data movement, but it has some limitations. For example, it does not support web service protocols (SOAP) and the overhead for maintaining the socket connection cannot be ignored. Instead, Reliable File Transfer (RFT) is a Web Service compliant component. It can be considered as a job scheduler for data movement. RTF provides users with an interactivation function throughout the data movement. Users should only give the URL list of resources. RTF edits the job description file according to the user's request, and can also answer the queries for transfer status as well as notify about the change of state.

Data Replica Components: Replica Location Service (RLS) and Data Replication Service (DRS)

RLS is a tool to trace the replica of data. The RLS server contains all the registries of the original data and the data replicas. Once a file is created, it is registered to a RSL server. Users send a request with the logical file name (the unique identifier of the original file) to the RLS server. The RLS server returns all the physical file names (the location of a copy of the original file) registered to it. The RLS server can be configured to be either distributed or centralized.

DRS is a WS component for higher level data management service. It works on top of RFT and RLS. It accepts the user's request for locating, transferring and registering new replicas of data files. It uses RFT for file transfer and RLS for registration and location mappings. In addition, DRS performs the data discovery function.

A.4.3 Job Submission (GRAM)

The GT4 component *Grid Resource Allocation Management (GRAM)* is a uniform interface for different resource management systems. It lies at the bottom level in the component architecture of GT. It can interact and control a variety of local schedulers such as CONDOR, PBS, LSF, etc., which are installed at different grid resource sites. GRAM processes the requests for resources needed for remote application execution, allocates the required resources and manages the active jobs. It has two implementations, *pre-WS GRAM* and *WS GRAM*.

In Pre-WS GRAM, the non-Web Service GRAM, the main components are the gatekeeper installed on a remote computer and the job manager, which is created by the gatekeeper. When a job is submitted, the gatekeeper of remote computer receives a request. The gatekeeper creates a job manager for the job to start and monitor this job. After the job is finished, the job manager also terminates.

WS GRAM is composed of a set of WSRF-compliant Web Services to locate, submit, monitor and cancel jobs on grid computing resources. These jobs can either be interactive jobs or jobs managed by a local scheduler. Jobs are started by the *ManagedExecutionJobService*, a Java service implementation running within the globus container. This Web Service accomplishes the function similar to that of a gatekeeper in the pre-WS GRAM. *ManagedJobFactoryService* creates an instance of stateful *ManagedJobService* for controlling and monitoring the job. It also publishes information about the computational resources. *ManagedJobService* is created by *ManagedJobFactoryService* and provides an interface to monitor the status of job, terminate the job, etc. The two data components, RFT and GridFTP, are used by WS GRAM. RFT is used for file staging before and after the job. GridFTP is required to access remote storage elements and file systems. It also monitors the status of file transfer.

A.4.4 Information Discovery (MDS)

Monitoring and Discovery System (MDS) is a suite of Web services to monitor and discover the grid resources and grid services available within a VO. This service module offers: Index service, Trigger service, WebMDS (Web Service Data Browser and underlying framework), and Aggregator Framework.

Collecting Information: Aggregator Framework

Aggregator Framework is the foundation of Index Service and Trigger Service. It is responsible for the information collection. In order to collect the resource information, Aggregator Framework utilizes the information systems at the lower level. Some of these information systems are GT components, such as WS GRAM and RFT. Others are the third party information systems. These information systems furnish:

- Information about the resources themselves, called resource properties.

- Information about the local scheduler (published by WS GRAM).

- Information concerned with file transfer such as the status of the server, transfer status and the number of active transfers (published by RFT).

- Information to identify which VO the services serve.

- Resource information published by other WSRF services.

Two Information Services: Index Service and Trigger Service

Index Service

Index Service has two functions. The first is the collection of monitoring and discovery information from grid resources. Its second is publishing the information as a service group at a single location. These functions are realized at a *VO* level. The Index Service offers two interfaces. One interface is used to put information into the index. The other is used to retrieve information from the index.

Trigger Service

Trigger Service also has two functions. It collects information about the grid resources and triggers some predefined action, if the collected data meets a specified criteria. The predefined actions may be sending a message to an end-user or writing to a structured log file.

Standard Web Interface: WebMDS

WebMDS is a standard web browser interface for users to view formatted information about grid services and resources. This component is implemented as a servlet and thus needs the support of a servlet-compliant web server. Tomcat is used as the web server for WebMDS. This component also offers a command-line tool (`webmds-create-context-file`) to configure Tomcat in order to automatically deploy WebMDS.

A.5   Using Globus

A.5.1   Definition of Job

A job can be defined as a process or set of processes created as an outcome of a job request.

A.5.2   Staging Files

File staging allows executables and data files to be automatically transferred to the required destination without user intervention. For file staging, specific elements have to be added to the provided job description XML file during the job submission. Each file transmission must provide a URL source and a URL destination. The URLs for remote files are specified as GridFTP URLs, and for the files local to the service they are mentioned as file URLs. The file URLs are internally converted to GridFTP URLs by the service. We now give an example of staging, that is transferring the file from the node that has submitted the job to a remote node where job execution will take place.

```
<fileStageIn>
    <transfer>
```

```
        <sourceURL>gsiftp://host1.examplegrid.org:2811/home/
            ...user1/userDataFile</sourceURL>
        <destinationURL>file:///${GLOBUS_USER_HOME}/
            ...transferred_files<destinationURL>
    </transfer>
</fileStageIn>
```

GLOBUS_USER_HOME refers to the home directory of the user on the remote host. So in this example the userDataFile is transferred to \~/transferred_files directory on the remote host.

In the job description file we can specify where the standard output and error files are to be directed. These files can then be staged *out* from the remote server. To redirect standard output and standard error we add the following to the job description file

```
<stdout>${GLOBUS_USER_HOME}/test.out<\stdout>
<stderr>${GLOBUS_USER_HOME}/test.err<\stderr>
```

After the completion of the job, these files can be transferred to the submission node by adding the following to the job description file.

```
<fileStageOut>
    <transfer>
        <sourceURL>file:///${GLOBUS_USER_HOME}/test.out<\
            ...sourceURL>
        <destinationURL>gsiftp://host1.examplegrid.org:2811/
            ...home/user</destinationURL>
    </transfer>
    <transfer>
        <sourceURL>file:///${GLOBUS_USER_HOME}/test.err<\
            ...sourceURL>
        <destinationURL>gsiftp://host1.examplegrid.org:2811/
            ...home/user</destinationURL>
    </transfer>
</fileStageOut>
```

Job description file can be used to automatically clean up the files transmitted to the remote node. Suppose we want to remove the file userDataFile from the remote host; we add the following lines to the job description file.

```
<fileCleanUp>
    <deletion>
        <file>file:///${GLOBUS_USER_HOME}/transferred_files/
            ...userDataFile</file>
    </deletion>
</fileCleanUp>
```

A.5.3   Job Submission

Data Transfer

Globus provides two ways for reliable and secure file transfer during remote job execution. Files can be transferred using the GASS protocol or the

GridFTP protocol. When using the GASS server, you have the option to choose between a secure or insecure communication channel. For secure connections, GSI is used for authentication. It uses proxy credentials generated by the command `grid-proxy-init`. The GASS protocol should be used when you need to exchange small files or support real-time information exchange. For large files you should use the GridFTP protocol. A GASS server can be started using the following command

```
globus-gass-server
```

This command supports a variety of options. The important ones are:

[`-w,-r`] These options allows a write access and a read access on the local filesystem.

[`-o,-e`] These options specify that output from the standard output (`<gass_server>/dev/output`) and the standard error (`<gass_server>/dev/error`) are redirected to the local output and the local error, i.e., the node that issues the job submission request and runs the GASS server.

[`-c`] This option enables the GASS client to shutdown the GASS server by writing to the logical file `<gass_server>/globus_gass_client_shutdown`.

In the above discussion, `<gass_server>` refers to the GASS server URL. Once the GASS server has been started, it can be used to transfer files based on the job description file.

When the files to be transferred are large, the GridFTP protocol should be used. GridFTP provides a command `globus-url-copy` to transfer files between two nodes. Like the GASS protocol, before using this command you must have a valid proxy certificate, generated by using the command `grid-proxy-init`. Then to start the GridFTP server execute the following command

```
globus-gridftp-server -p 2811 -S
```

This starts the server on the default port (2811). The `-S` option starts the server in the background. In description of `globus-url-copy`, we use the terminology used in FTP for file transfers. To put a file on the GridFTP server, we issue the following command

```
globus-url-copy -vb file:///home/user1/temp gsiftp://host1.
    ...examplergid.org/usr/local/temp
```

This command transfers the file temp in the user1's home directory to a file named `/usr/local/temp` on the server running at `host1.examplegrid.org`. To get the file from the server, you just reverse the source and destination URLs. Now suppose, you want to get the file just copied to the directory `/tmp` as file `foo`, you simply issue the following command

```
globus-url-copy -vb gsiftp://host1.examplegrid.org/usr/local/
    ...temp file:///tmp/foo
```

For a complete list of options and URL prefixes supported by globus-url-copy, you can refer to the online documentation available at `http://www.globus.org/toolkit/docs/4.0/data/gridftp/GridFTP_Commandline_Frag.html`.

Job Submission

Globus Resource Allocation Manager (GRAM) provides a set of client tools for submitting jobs remotely. In this section we look at those tools and the possible options. `globus-job-run` provides a job-submission interface and allows staging of data and executable files using the GASS server. The command can be executed as

```
globus-job-run machine_name:port/jobmanager-name command
```

The default port number is 2119 and the default jobmanager-name is `jobmanager`. Suppose you have a jobmanager running by the name `jobmanager-date` on the localhost. You would execute

```
globus-job-run localhost:2119/jobmanager-date /bin/date
```

Using the -s flag with the command gives access to the staging functionality by starting a GASS server on the localhost. Suppose you have an executable by the name `exampleProg`. To stage the file, you would execute the following command

```
globus-job-run localhost -s exampleProg
```

This will start a GASS server on your machine. The jobmanager will contact the GASS server to read the staged file and finally submit the job to the scheduler.

`globus-job-submit` is similar to `globus-job-run` except that it provides a batch interface. Currently this command does not support automatic staging of files. The syntax of this command is same as `globus-job-run`. `globus-job-submit` returns immediately providing a contact string, which you can use to query the status of your job.

The `globusrun` command gives access to the complete features of the *Resource Specification Language (RSL)*. You can provide an RSL file by specifying a -f option to the `globusrun` command.

```
globusrun -r host1.examplegrid.org/jobmanager-pbs -f example.
    ...rsl
```

The `example.rsl` file contains the complete job description including the executable file, input and output files and the details about file staging. The `globus-job-run` and `globus-job-submit` commands are shell script wrappers around the `globusrun` command.

The GRAM component of globus toolkit supports two approaches: a Web Service GRAM approach and pre-Web Service GRAM approach. The commands for job submission described in this section are the job submission clients that use the pre-WS GRAM approach. The commands based on the WS GRAM approach have a similar syntax, but are followed by a `-ws` suffix. For example the `globusrun` command becomes `globusrun-ws` and so on.

A.5.4   Job Monitoring

Finding Status of Submitted Jobs

The `globus-job-status` displays the status of job submitted by the command `globus-job-submit`. The `globus-job-submit` returns a 'contact string', which can be used to query the status of the submitted job. The contact string returned is of the form `http://host2.examplegrid.org:5678/12340/1176891`. The states of a submitted job can be one of the following: *Unsubmitted*, *StageIn*, *Pending*, *Active*, *Suspended*, *StageOut*, *Done* or *Failed*.

Collecting Output and Cleaning Files

The contact string returned by the `globus-job-submit` command, can also be used to get the output of the executed job. To get the output of the job, run the following command

```
globus -job -get -output 'contact string'
```

The `globus-job-clean` command cleans up the cached copy of the output generated by the job. If the job is running while this command is executed, the running job is stopped.

References

[171] Globus Alliance. Globus v4.0.5. Web Published, 2007. Available online at: `http://www.globus.org/` (accessed January 1st, 2009).

[172] Borja Sotomayor. The Globus Toolkit 4: programmers tutorial. Web Published, 2005. Available online at: `http://gdp.globus.org/gt4-tutorial/` (accessed January 1st, 2009).

[173] Globus Alliance. GT4 admin guide. Web Published, 2007. Available online at: `http://www.globus.org/toolkit/docs/4.0/admin/docbook/` (accessed January 1st, 2009).

[174] Globus Alliance. GT 4.0.x quickstart guide. Web Published, 2005. Available online at: `http://www.globus.org/toolkit/docs/4.0/admin/docbook/quickstart.html` (accessed January 1st, 2009).

[175] Globus Alliance. Globus Toolkit 4.0: release manuals. Web Published, 2005. Available online at: `http://www.globus.org/toolkit/docs/4.0/` (accessed January 1st, 2009).

Appendix B

gLite

B.1   Introduction

gLite is grid middleware developed by the Enabling Grids for E-sciencE (EGEE) project to provide a set of tools for resource sharing. It combines the collaborative efforts of scientists and engineers from over 32 countries worldwide to create a grid infrastructure that is robust and highly accessible. By collaborating with 30,000 CPUs and about 5 petabytes of storage, gLite is middleware that helps to fulfill the demands of huge computational power in various scientific fields of geology, finance, physics and climatology.

A new trend to migrate grid applications from traditional scientific areas such as high energy physics and life sciences to other domains including computational finance and geology is gaining interest. This trend means that grid infrastructures is going to change scientific research drastically in the near future. It is not unlikely that grid middleware such as gLite is going to be used for day-to-day operations of people. Nonetheless, gLite is a young project and still has a long way to go before it reaches maturity and a level of stability suitable for widespread public usage.

In this chapter, we present a brief introduction of the main components of gLite. The internal workings of the components are also highlighted to give a better comprehension of the middleware architecture. The key components of gLite are as follows:

- Information and discovery service

- Security

- Computational and storage elements

- Logging and bookkeeping services

- Workload management service

The usage of gLite for the submission, retrieval, query and cancelation of jobs is also demonstrated to give readers the basic tools required to run gLite. Commands and examples of job description language for sequential and parallel jobs are also presented to give simple and direct hands-on experience to the reader.

B.2   Internal Workings of gLite

B.2.1   Information Service

This service provides information on the status of the resources, which is crucial for the operations on the grid. It is also through this service that resources are discovered and monitored. There are two information service mechanisms in *gLite*. They are namely *relational grid monitoring architecture (R-GMA)* for accounting, monitoring and publication of user level information, and *globus monitoring and discovery service (MDS)*, which is used for resource discovery and for resource status publication. The functions of information service in *gLite* are presented in Figure B.1.

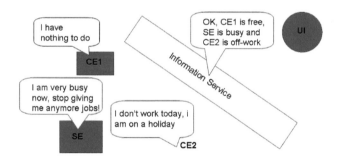

FIGURE B.1: Functions of information service in *gLite*.

Relational Grid Monitoring Architecture (R-GMA)

R-GMA Model

Based on the Grid Monitoring Architecture (GMA), R-GMA is a relational database that facilitates the query of user information. It is a structure that comprises three basic components, namely registry, producers and consumers. The producers publish the resources that they have to offer and the consumers collect resource information that they need from the tables in the database. In the R-GMA model, the database is a virtual one in the sense that it uses a registry to mediate between producers and consumers rather than having an actual database. Figure B.2 gives a basic representation of the R-GMA model between producers, consumers and registry.

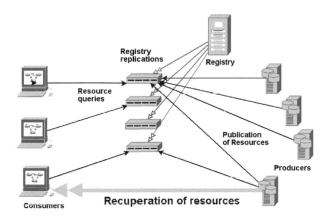

FIGURE B.2: The replication of registry tables for multiple access to resources in the producers.

Servlet Technology

The R-GMA model employs the servlet technology for the inter-communication interface between consumers, producers and registry in *gLite*. By establishing a communication protocol, consumers, registry and producers are able to transmit information in an efficient way among each other. The use of this technology also allows for easy and rapid adaptation to other web service systems. The R-GMA model is currently used for the monitoring of system and user level information. It contains the same information as the BDII server, but it is not currently used for the submission of jobs in *workload manager* in *gLite*.

Globus Monitoring and Discovering System (MDS)

MDS Architecture

The MDS essentially possesses a tree structure where information about resources are gathered at each level of the tree as illustrated in Figure B.3.

Grid Resource Information Server (GRIS)

We can see from Figure B.3 that information about individual resources at the fundamental level are generated by the information provider that can

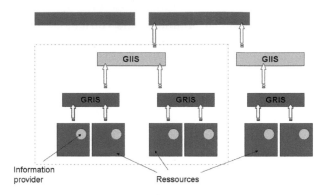

FIGURE B.3: Hierarchic tree structure of monitoring and discovering system.

be found on most computing and storage elements. This information is then collected and broadcasted by the *grid resource information server (GRIS)*, which is essentially a *lightweight directory access protocol (LDAP)* server. It is also important to remark that the GRIS server on each resource element (CE or SE) manages all information pertinent to a particular resource element. This information is then transmitted to the next level, which is the GISS server.

Site Grid Index Information Server (GRIIS)

GRISS is also an LDAP server responsible for the publication of all resource information on the same site. This information is obtained by congregating information from every GRIS on the site.

Berkeley Database Information Index server (BDII)

BDII is the LDAP server at the top level of the MDS structure. Each BDII server is configured to monitor a certain number of GRIIS where resource information about a particular site is transmitted for querying in the grid. As BDIIs are at the highest level in MDS, BDII contains all the information about the resources available at the sites that it is monitoring. Lastly, at the highest level of all BDIIs in a *virtual organization (VO)*, there is an overall BDII that helps to link up the various BDII in the same VO. This is a server where all useful information concerning the VO is stored. Information service is a crucial component in the structure of *gLite* as it plays an important role in the monitoring and discovering of resources essential for the normal operation of gLite.

B.2.2 Workload Management System

The *workload management system* is the main component in the *gLite* structure. It manages the submission and cancellation of jobs from the user interface via the *network service*. It is, in general, a mediating service that helps
to schedule jobs into *gLite* in the most efficient way by soliciting important
information on the status of the resources available at any one time.

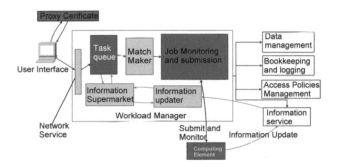

FIGURE B.4: Internal structure of work management service.

Overview of Workload Manager

The *workload manager (WM)* is the principal component in the WMS. It
comprises task queue, matchmaker, information supermarket, information updater and job handler. It's principal role involves user level job requests by
carrying out optimal job allocation and packaging before submission to the
resource elements. The job requests are written primarily in a *job description
language (JDL)*. This process of allocation and packaging is done with input
from other grid services, namely *data management*, *bookkeeping* and *information services* of the grid. It is through the workload manager that status of
submitted jobs retrieved from the information service are relayed to the user
interface.

Task Queue

This is the component in *workload manager* that deals with the handling
of a series of job requests from the user interface via the *network service*.
The network service acts as the portal for accepting all external requests to
the workload manager. It also carries out the authentication of user for the
utilization of grid resources via digital proxy certificates. Once the job request
(job submission or job cancellation) is submitted to the *network service*, the
request in JDL is interpreted before passing the request to the task queue.
Once the job request reaches the task queue, it is enqueued into the job

request list for an eventual execution of the request upon scheduling by the matchmaker. It is important to note that the passing of job requests to the matchmaker is a two-way process. If no suitable resource is available for scheduling, the job request is returned to the task queue for eventual dispatching to the matchmaker later on.

Matchmaker

Matchmaker is the component in *workload manager* where requests are processed to match the most suitable resource available. The matchmaking is done in two parts, namely the processing of job request, and the solicitation of resource information. On one hand, the processing of job requests is based on several criteria, which include access policies, user job preferences and nature of the job requested. On the other hand, the solicitation of resource information is carried out by querying the information supermarket, which contains up-to-date information regarding the resources in the grid. When both the parts of matchmaker are ready, the matchmaking process can be carried out to optimize the scheduling of job requests to grid resources available. The processed job request is passed onto the next stage for actual job submission to the resource elements.

Information Supermarket

The information supermarket is basically a cache where overall resource information about the grid is stored for querying by the matchmaker for matchmaking decisions. The information stored in the information supermarket is constantly updated via the information updater.

Information Updater

The information updater is a component where the updating of resource information in the information supermarket is done. This updating of information is carried out by reading information from the top-level BDII server of the virtual organization.

Job Handler

When the process of matchmaking is done, the job request is submitted to the resource elements for processing by the job handler. The job handler consists of four basic components that are responsible for: job packaging, job submission and removal and job monitoring. The process of job packaging is done by the job adapter and the job controller. In the job adapter, the JDL is modified for an eventual submission by the job controller. The job wrapper script is also written in the job adapter for creating a suitable execution environment into the *computing element*. Once the JDL has been processed by the job adapter, it is passed onto the job controller where the job is submitted by *CondorC*. *CondorC* is the primary component that is responsible for performing actual job submissions and removals. The submitted jobs are then actively monitored by the *log monitor* for appropriate actions in response to job state events.

B.2.3 Job Description Language (JDL)

The JDL is a high-level language used for parameterizing a job with specific user parameters. The JDL is used by the WMS to understand the specification of the job in order to find the best-matching grid resource. Some of the basics of JDL are exposed in this section but it is by no means the complete JDL guide. Nonetheless, users should be able to launch a basic job using the description in this section.

Principles

The JDL file is a job description that is essentially made up of lines having a format like

```
attribute = expression
```

The expressions can be of variable length but it must be terminated by a semi-colon. Moreover, if double quotes are present, they must be prefixed by a backslash (\") since the apostrophe character is invalid in JDL. Comments can be preceded by // or a \# character. For multi-line comments, /* and */ can be used. It is important to note that JDL is sensitive to blank spaces, and blank spaces following the semi-colon would render the JDL invalid.

To Run a Simple Job

To execute a program on the grid, we can write a JDL file in the following way.

```
[
JobType =   "Normal";
RetryCount = 0;
ShallowRetryCount = 3;
Executable =  "compress\_hgdemo.sh";
Arguments =  "argA argB";
InputSandbox =  {"file:///home/compress_hgdemo.sh"};
StdOutput = "std.out";
StdError =  "std.err";
OutputSandbox = {"std.out", "std.err","fileoutput"};
DataAccessProtocol =  {"gsiftp", "https"};
]
```

The `JobType` defines the type of job to be submitted. It can be sequential in the case of ''Normal'' and parallel jobs in the case of ''Parametric''.

The `RetryCount` and `ShallowRetryCount` allow for the auto-submission of jobs until the job is submitted. These attributes specify the maximum number of times a job is resubmitted before it is aborted. If it is not defined, it takes the default values in the WMS defined by `MaxRetryCount` and `MaxShallowRetryCount`. `RetryCount` corresponds to deep resubmission while `ShallowRetryCount` corresponds to shallow resubmission. A resubmission is deep when the resubmitted job is a job that has failed after it has started execution.

The `Executable` attribute specifies the command to be run by a job. Only the filename of the executable is required.

The `InputSandbox` is the attribute for storing the command path on the UI so that the grid system knows where to search for the executable file precisely. If other files have to be used for the executable, e.g., the arguments, they have to be specified in the `InputSandbox`. The files in the `InputSandbox` must not have the same name, otherwise they overwrite each other during the file transfer. `Arguments=``Numtest3'';` can be specified if it is required by the executable file.

The attributes `StdOutput` and `StdError` define the name of files that contains the standard output and standard error once the job is completed.

The `OutputSandbox` contains files that are returned by the grid once the execution of the job is finished. Normally, it contains the files ``std.out'' and ``std.err''.

The `DataAccessProtocol` defines the protocol used for the transfer of the files and the retrieval of the job outputs.

To Run a Parallel Job

To execute a parallel job, we can write the JDL file in the following way.

```
[
JobType =   "Parametric";
Parameters = (N) ;
ParameterStart = 1;
ParameterStep = 1;
RetryCount = 0;
ShallowRetryCount = 3;
Executable = "compress\_hgdemo.sh";
Arguments = "argA_PARAM_ argB_PARAM_";
InputSandbox = {"file:///home/compress_hgdemo.sh"};
StdOutput = "std.out";
StdError = "std.err";
OutputSandbox = {"std.out", "std.err","fileoutput"};
DataAccessProtocol = {"gsiftp", "https"};
]
```

The only difference between sequential jobs and parallel jobs lies in the definition of the JDL file. In parallel jobs, the JDL file has three additional attributes: `Parameters`, `ParameterStart` and `ParameterStep`.

First the `jobtype` has to be defined as ``Parametric''. Thereafter, `Parameters` define the maximum number for the attribute `\_PARAM\_`, while `ParameterStart` defines the starting number for `\_PARAM\_` and `ParameterStep` defines the increment of `\_PARAM\_`. With these three attributes, we are able to define `\_PARAM\_` as a series of numbers. This is important as *gLite* uses these numbers as parameters to the arguments and run jobs in parallel.

For example, if `\_PARAM\_` is a series $1, 2, 3, 4, \ldots, 10$, the above example requires `argA1, argA2, ..., argA10` for the jobs. Each argument is supplied to a computing node and each job is run in parallel. It is important to note that the executable file remains the same and that only the parameters are changed. In other words, the jobs are run in parallel using the same executable file with different parameters for each job.

B.2.4   Computing Element

Computing element is a cluster of grid resources that is used for running the jobs. Each computing element is composed of three basic elements, notably *grid gate*, *local resource management system* and *worker nodes*.

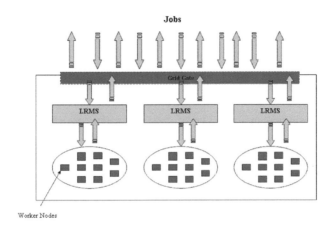

FIGURE B.5: Structure of computing element.

Grid Gate

Grid gate is the interface between the cluster and the grid. It regulates the input and output of the jobs in the computing element. Common implementations include *LCG grid element* and *gLite* grid element.

Allocation of Resources (R-GMA)

Also commonly noted as the local resource management service, this service manages the allocation of local resources in the cluster. Common implementations include *Maui/Torque* and *Condor*.

Worker Nodes

Worker nodes are generally the processors where the jobs are actually run. The strict definition defines the worker nodes as a component of the computing element, which is a queue in the LRMS. Thus, different queues in the same cluster correspond to different computing elements.

B.2.5   Data Management

Data management in *gLite* can be grouped under three categories: *storage elements*, *file catalogues* and *file transfer service*. These three components are essential for data access in *gLite*.

Storage Elements (SE)

SE are components in *gLite* that store data for eventual retrieval by user or other applications in *gLite*. In a grid environment, files are replicated and stored in different locations to provide rapid multi-access capabilities. Thus, in order to ensure the consistency and therefore the uniqueness of the data, the files are limited to read-only properties. In this write-once-read-many environment, files are written once during creation and cannot be modified thereafter.

To define a storage element, we need to know the *storage resource manager*, *storage resource type* and the *transfer protocol* used by the storage resource manager.

Storage Resource Manager (SMR)

The storage resource manager is a middleware interface application that renders standard data management operations between SEs of different resource type transparent to user. These data management operations include file transfer, space reservation, renaming of files and file directory creation.

Storage Resource Types

There are a number of storage resource types available but they can be broken down into two basic categories, namely disk-based and tape-based storage. For relatively small SEs, disk-based storage implementation is employed together with disk pool manager, which is the storage resource manager corresponding to disk-based storage. For bigger SEs, the *mass storage system (MSS)* is implemented with *CASTOR (CERN Advanced STORage Manager)* as the storage resource manager. For hybrids between disk pool storage and MSS, we have *dCache* as the storage resource manager.

Transfer Protocol

Based on the resource type of the SEs, different protocols are used for the transfer of files in and out of the SE. For file transfer with SEs, *grid security infrastructure file transfer protocol (GSIFTP)* is used to facilitate fast and secure transfer of data. For remote access to files on SEs, the *remote file input/output protocol (RFIO)* is used. Thus, based on the storage resource type in each SE, we have a corresponding storage resource manager that supports protocols for different actions to the SEs.

Local File Catalog

Before normal operations of transfer and access can be performed on files in the SEs, they have to be located and identified. This identification of files on the storage elements is done through the use of different identifiers, namely *Grid Unique Identifier (GUID), logical file name (LFN), Storage URL (SURL)* and *Transport URL (TURL)*.

Grid Unique Identifier (GUID)

The GUID is an identifier that uniquely identifies the files in the grid. It is associated with a file when it is first created using a combination of MAC address and timestamp.

Logical File Name (LFN)

Although GUID identifies uniquely a file in the grid, they are not commonly used for locating a file. Instead, the LFN is used. It is an identifier that utilizes intuitive human readable strings as the identifier. This in turn facilitates interaction between user and the grid. It is also important to note that the LFN of a file is not unique and a file can have many LFNs for multiple references.

Storage URL (SURL)

Also known as *physical file name (PFN)*, the storage URL of a file contains the physical address of the file replicated in the grid. It provides information of the SE or its SRM and the path needed to locate the files on the SE involved.

Transport URL (TURL)

While storage URL provides location information of the file in the grid, transport URL provides access and retrieval information on the file. On first look, it may seem redundant that transport URL may contain identical information as storage URL. However, there is more than meets the eye. The transport URL contains also access protocol and physical port information that allow for the physical retrieval of the file concerned. This information is generated by the SRMs and can change over time.

Thus, several transport URLs for a given file exist and the number of transport URLs corresponds to the number of protocols supported by the storage resource manager. Furthermore, the storage resource manager may hold several duplicated files to reduce the bottleneck in accessing the files.

It is important to notice that while GUIDs and LFNs are used for the identification of files, storage URLs and transport URLs provide the necessary information to access and retrieve the files concerned.

Lastly, the GUIDs, LFNs, storage URL and transport URL are registered in the *Local File Catalog (LFC)* where mappings between file identifiers are stored. This provides a source of reference to locate and access files in the grid. An analogy of a LFC would be the address book that contains contact information of a list of people. Each person has a name and a residential address where we can physically locate the person. In this analogy, the name would be comparable to GUIDs and LFNs while the residential address is comparable to the storage and transport URLs. When the grid wants to obtain information about a particular file, it queries the LFC in order to obtain access and retrieval information of the file involved.

File Transfer Service (FTS)

File transfer service (FTS) is a low-level service that permits the transfer of data between different SEs. In general, the transfer protocol used between SEs is the *Grid Security Infrastructure File Transfer Protocol (GSIFTP)*. In fact the protocol, needed for a particular transfer, is established based on the protocols supported by the storage resource managers in both the source and destination SEs. We now introduce some of the terminologies used in FTS.

- Channel: this is a specific pathway for the transfer of files. While production channels ensure a minimum throughput, non-production channels are usually open networks that do not ensure such a minimum throughput.

- File: a set of source and destination storage URLs.

- Transfer Job: a set of files with additional parameters including cryptography of data transferred.

- File State: the state of an individual file transfer.

- Job State: the state of the transferred job based on file states of the underlying files.

Once the transfer job has been submitted, it returns a job ID, which permits query of the progress of the job state or cancellation of the job. It is important to outline that FTS presently supports only storage URL pairs instead of GUID/LFN pairs, which are more intuitive. The transfer job states of a job are as indicated in Figure B.6.

B.3 Logging and Book-Keeping (LB)

The *Logging and Book-keeping (LB)* module allows for the tracking of major job events: *matchmaking*, job submission, job cancelation, etc. This logged

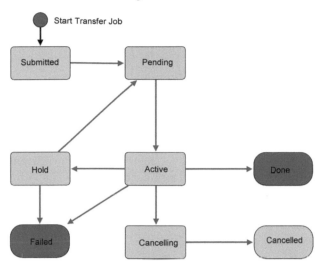

FIGURE B.6: Transition between transfer job states.

information is obtained from the WMS or the CEs. This information is then passed to a nearby LB component (*locallogger*) in order to avoid network problems. The locallogger has the responsibility of passing this information to the main LB component named the Book-keeping Server.

At the Book-keeping Server, the job-event information is then processed and translated into higher-level job information (SUBMITTED, RUNNING, DONE, etc). This Book-keeping Server is assigned to a job for the entire lifetime of the job. Query of job events can be done using the UI.

The LB module also provides the possibility to notify users when there are particular changes in the job. This is carried out systematically in the grid architecture where jobs are given a unique URL job id. This coincides with the URL of the Book-keeping Server.

The tracking of job events is done through close monitoring of the WMS where job events are reported back from the grid resources. A locallogger is situated near the WMS to avoid network problems. This locallogger then transmits the information to the central Book-keeping Server corresponding to the job. This transmission of information is carried out by the *interloggers* to continuously pass the information until it reaches the Book-keeping Server.

B.4   Security Mechanism

In this section, we discuss the usage and the important role of certificates that serve as the passports in the world of grid. *Certificates* under the *Grid Security Infrastructure (GSI)* are important documents that ensure the authenticity and security of the data transferred between different grid components. It also allows for the inter-verification of each individual component in the grid. The basics in security infrastructure, already discussed in Chapter 4, are necessary to understand the following. In particular, the concepts of private and public key encryption, signature, and certificates are fundamental.

In the next paragraph, we briefly discuss the role of proxies in security and remote delegation of grid operations.

In *gLite*, we are often faced with the situation where a remote service acts on behalf of the user, which is called delegation. The user has to authorize the remote service before anything can be done. The authorization process is carried out through the use of a certificate that is signed by the private keys of the user (end user X.509 certificate). This certificate of delegation is call the *proxy*. With the proxy, jobs submitted by the remote service is identified as a delegation by the user and the proxy is queried with the user certificate to verify the delegation.

When users register for an account, they get a user name and a password. Using this information, an authorized user can ask for a year-long X.509 digital certificate that is saved in the home directory of the user. Upon receiving this certificate, users are now ready to carry out jobs on the grid under the Grid Security Infrastructure (GSI). However, before going onto the grid, users have to log in to the grid. This is done through the creation of temporary proxy certificates using the user certificate. The proxy certificate lasts for 12 hours. The duration of 12 hours is designed to reduce the impact of an eventual security compromise, e.g., theft of a proxy certificate.

gLite resources are also given certificates to ensure their authenticity and to enable identification of these resources. In this way, components in the grid can be identified and authenticated easily.

The identification and authorization of users by the resources is done through 2 mechanisms, notably by the gridmap-file mechanism and the *virtual organization membership service (VOMS)* mechanism. Upon authorized access to the resources, the user is able to perform basic grid operations.

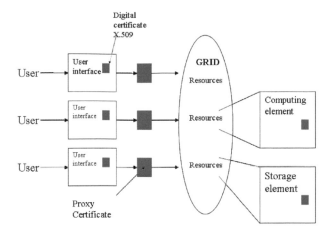

FIGURE B.7: Overview of certificates in the security mechanism in gLite.

B.5 Using gLite

In this section, we give an overview of the frequently used commands involved in the user interface throughout the life-cycle of a job in gLite. The principal phases pertinent to a user are the submission of jobs to gLite, the recuperation of job status and the collection of job output in gLite.

B.5.1 Initialization

Before any operations can be carried out in gLite, users must first create a proxy (similar to a certificate) as remarked in the previous section. This proxy has a duration of 12 hours during which users can carry out various operations pertinent to gLite. This initialization process is done through the command voms-proxy-init-voms<VO> where VO is the name of the virtual organization (VO). The grid password is required from the user before proxies can be attributed. The output is similar to the following:

```
Your identity: /C=CH/O=CERN/OU=GRID/CN=John Doe
Enter GRID pass phrase:
Creating temporary proxy
.......................................... Done
Contacting lcg-voms.cern.ch:15002
[/C=CH/O=CERN/OU=GRID/CN=host/lcg-voms.cern.ch]
"cms" Done
Creating proxy
.............................................. Done
Your proxy is valid until Thu Mar 30 06:17:27 2006
```

This provides a level of authentication of users before grid resources are made accessible to them. Once proxies are attributed to users, users can proceed to the next phase of utilization.

B.5.2 Job Paths: From Submission to Collection

We now illustrate a typical life-cycle from submission of jobs to the eventual collection of job outputs. The complete life-cycle is represented in Figure B.8.

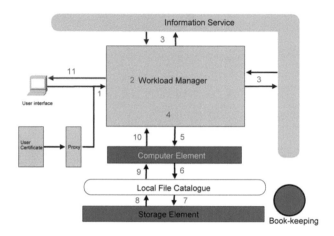

FIGURE B.8: Pathway followed by a job from submission to collection of job results.

The details of the operations are the following:

1. The job is submitted through the user interface with the proxy created for a duration of 12 hours. The job is authenticated with the proxy and the job enters the *Workload Manager*.

2. The job enters the *Workload Manager* and waits for the matchmaking process to find a suitable resource to process the job.

3. The *Workload Manager* queries the information service of the grid to look for suitable resources in the grid. Information on suitable resources will then be returned to the *Workload Manager*.

4. With suitable resources available, the job is packaged for actual submission to the computing element.

5. The job is submitted to the computing element for processing.

6. For job processing, the computing element requires files stored in the storage element. The location of the files is obtained by querying the local file catalog.

7. The resource on the storage element is located.

8. The location of the resource is confirmed.

9. The location of the resource is passed to the computing element. The resource (file) is retrieved for processing in the computing element.

10. The job is executed successfully and the output URL is sent to the *Workload Manager*

11. The output URL is passed on to the user interface where the user can retrieve the output of the job from the grid.

It is important to note that throughout the whole process from 1 to 11, the status of the job is logged by the Book-keeping service. In this way, the whole process is kept in record for eventual troubleshooting should the need arise.

B.5.3 Job Submission

This phase is a very important part of the job's life-cycle in gLite. It is in this phase that users specify the nature of the job (parallel or sequential) and other parameters in order to launch the job. These specifications are calibrated through the use of Job Description Language (JDL). The detailed syntax of JDL files can be found in Section B.2.3. As mentioned earlier, JDL files are used for the configuration of jobs in *gLite* as they allow *gLite* to understand the job parameters required by users at the user interface.

It is important to notice that the *gLite* WMS has two implementations for the management of jobs, namely *network server (NS)* and *Work Management Proxy (WMProxy)*. Both implementations offer similar services and the command line commands are highly similar. However, WMProxy offers more substantial gain and as follows:

- Submission of job collections

- Faster authentication

- Faster matchmaking

- Faster response time for users

- Higher job throughput

The submission of job is done through the use of the following commands: jdlfile is the file written in JDL specifying the parameters and configuration of the job while joblist is the file that contains the URL of the job, equivalent to the job ID.

If the job has been successfully submitted, *gLite* returns as output the job ID, which is URL by which the status of the job can be queried via the user interface. A typical output obtained after successful submission is as follows:

```
Connecting to the service...
https://rb102.cern.ch:7443/glite_wms_wmproxy_server
==================== glite-wms-job-submit Success ============
The job has been successfully submitted to the WMProxy Your
job identifier is:
https://rb102.cern.ch:9000/vZKKk3gdBla6RySximq_vQ
============================================================
```

If the command [-ojoblist] has been used, the job identifier (URL) will be stored in the file joblist. The list of job identifiers in the file joblist can then be retrieved in other job operations through the use of -i command.

Function	via WMproxy
To submit a job in *gLite*	glite-wms-job-submit[-ojoblist]-ajdlfile
Function	via NS
To submit a job in *gLite*	glite-job-submit[-ojoblist]-ajdlfile

Table B.1: Job submission commands depending on the services.

B.5.4   Retrieving Job Status

Once the job has been submitted to the WMS of gLite, job status can be obtained from the job history log by querying the Logging and Book-keeping service of gLite. Table B.2 gives a number of commonly used commands for the retrieval of job status. The command -i<filename> allows the retrieval of status of jobs where job ids are stored in the file with the name of <filename>. This usually coincides with the file used with the command -o in the submission of jobs.

If job status is retrieved correctly, the output is similar to the following:

```
******************************************************************
BOOKKEEPING INFORMATION: Status info for the Job:
https://rb102.cern.ch:9000/fNdD4FW_Xxkt2s2aZJeoeg
Current Status: Done (Success)
Exit code: 0
Status Reason: Job terminated successfully
Destination: ce1.inrne.bas.bg:2119/jobmanager-lcgpbs-cms
Submitted: Mon Dec 4 15:05:43 2006 CET
******************************************************************
```

Function	via WMproxy
To retrieve job status	`glite-wms-job-status''jobURL''`
To retrieve job status	`glite-wms-job-status-i<filename>`
Function	via NS
To retrieve job status	`glite-job-status''jobURL''`
To retrieve job status	`glite-job-status-i<filename>`

Table B.2: Job status retrieval commands depending on the services.

It is important to notice that users can also specify the verbosity of the status that they want. This is done through the command -v[num] where num is a number from 0 to 2. If no verbosity is specified, the default value is 0. As the verbosity increases, the information on job status becomes more and more detailed. For example, a verbosity level of 2 gives the timestamps of the various stages of a job (from Submitted to Done/ Canceled/ Aborted). This allows users to know, for instance, the duration of actually running a job in the Computing Element (CE).

B.5.5 Canceling a Job

If jobs submitted to WMS in *gLite* need to be canceled, the commands in Table B.3 are used.

Function	via WMproxy
To cancel a job	`glite-wms-job-cancel<jobID>`
To cancel a job	`glite-wms-job-cancel-i<filename>`
Function	via NS
To cancel a job	`glite-job-cancel<jobID>`
To cancel a job	`glite-job-cancel-i<filename>`

Table B.3: Job cancelation commands depending on the services.

Similar to use of job status, -i command allows users to choose from a list of jobs that they want to cancel. If the cancellation of jobs is successful, users will get a similar output as follows:

```
Connecting to the service
https://128.142.160.93:7443/glite_wms_wmproxy_server
```

```
================== glite-wms-job-cancel Success ============
The cancellation request has been successfully submitted for
the following job(s): -
https://rb102.cern.ch:9000/P1c6ORFsrIZ9mnBALa7yZA
============================================================
```

B.5.6   Collecting Results of a Job

Once the job status has reached the status DONE, the output of the completed job can be retrieved.

Function	via WMproxy
To retrieve a job output	`glite-wms-job-output<jobID>`
To retrieve a job output	`glite-wms-job-output-i<filename>`
Function	via NS
To retrieve a job output	`glite-job-output<jobID>`
To retrieve a job output	`glite-job-output-i<filename>`

Table B.4: Job retrieval commands depending on the services.

Similar to the above, -i command allows users to choose from a list of jobs from which the output is retrieved. If the output of a job is successfully retrieved, an output similar to the following will be obtained:

```
Connecting to the service
https://128.142.160.93:7443/glite_wms_wmproxy_server
============================================================
JOB GET OUTPUT OUTCOME Output sandbox files for the job:
https://rb102.cern.ch:9000/yabp72aERhofLA6W2-LrJw have been
successfully retrieved and stored in the directory:
/tmp/doe_yabp72aERhofLA6W2-LrJw
============================================================
```

Once the output directory of the job is known, the results of a job can be accessed and the output can be retrieved by the user. It is important to note that job output directories are temporary and they are deleted after a certain amount of time depending on the administrator of the WMS.

References

[176] Stephen Burke, Simone Campana, Antonio Delgados Peris, Flavia Donno, Patricia Mendez Lorenzo, Roberto Santinelli, and Andrea Sciaba. *gLite 3: user guide*. Worldwide LHC computing Grid, January 2007.

[177] Xavier Jeannin. Vue d'ensemble de l'architecture logicielle. Web Published, October 2006. Available online at: http://www.urec.cnrs.fr/IMG/pdf/JTR2006-architecture_gLite.pdf (accessed January 1st, 2009).

[178] EGEE. Running on the grid using gLite 1.3. Web Published, August 2005. Available online at: http://egee-jra1.web.cern.ch/egee-jra1/Documentation/Tutorial/EGEE-JRA1-TEC-584530-GLITETUTORIAL-v0-1.pdf (accessed January 1st, 2009).

[179] EGEE. R-GMA command line tool. Web Published, November 2005. Available online at: http://hepunx.rl.ac.uk/egee/jra1-uk/glite-r1.5/command-line.pdf (accessed January 1st, 2009).

[180] Sergio Andreozzi. The gLite middleware. Web Published, February 2007. Available online at: http://omii-europe.forge.cnaf.infn.it/_media/jra2/documentation/djra2.0/glite-training.ppt (accessed January 1st, 2009).

[181] EGEE. gLite data management. Web Published, January 2007. Available online at: http://www.apan.net/meetings/manila2007/presentations/egee/gLite_datamanagement.ppt (accessed January 1st, 2009).

[182] EGEE. gLite installation guide. Web Published, 2007. Available online at: http://glite.web.cern.ch/glite/packages/R1.0/R20050331/doc/installation_guide.html (accessed January 1st, 2009).

Appendix C

Advanced Installation of gLite

The details of the installation and the configuration of gLite are presented to provide the reader with a basic guide for gLite installation and gLite configuration. Although this chapter helps the user to better understand the internal mechanisms of gLite, the interested reader is referred to the gLite manual for a complete understanding.

C.1 Installation Overview

The purpose of this section is to give a very brief overview for the installation of *gLite*. Through this section, the readers are able to understand the basic mechanism and structure of the installation procedure in deep details.

C.1.1 Deployment of gLite

gLite is service-oriented grid middleware whose architecture is basically composed of services used for data and resource management. These services are normally installed on dedicated computers. However, *gLite* also gives the ability to install various services on the same computer when computational resources are limited. The services that are to be installed for *gLite* are as follows:

- Security

- *Computing Element* and *Worker Node*

- *Workload Manager*

- *Logging and Book-keeping Server*

- *gLite I/O*

- *Local Transfer Service*

- *Single Catalog*

- Information and monitoring system

- VOMS server and administrator tools

- User Interface

C.1.2   gLite Packages Download and Configuration

gLite RPM packages for various services can be downloaded and installed based on installation scripts that come with the package. These packages can be obtained from `http://glite.web.cern.ch/glite/packages`. A configuration script is also provided for the configuration, deployment and the starting of services in *gLite*. These scripts can be executed once the RPM packages are installed to help in the configuration of the environment parameters. It is important to note that the configuration and installation of the services are kept separate to provide flexibility in the configuration of the service based on the requirements and constraints of the site. The details of the configuration parameters will not be presented here as it is beyond the scope of this book.

Installation Prerequisites

The following installation procedures are based on the assumption that the server platform is Red-Hat Linux 3.0 or any other compatible distribution such as Scientific Linux or CentOS. For operations concerning the components of *gLite*, the Java Development Kit (JDK) or Java Resource Environment (JRE) (version 1.4.2 or higher) is required. The version of Java can be configured in the `glite-global-cfg.xml` file. The parameters in this file have to be updated to the current version used in the installation.

During Installation

For the installation of each component except UI, the first three steps are always the same except for changing some variables. Thus, in order to lighten the literature here, we give a generic set of instructions for these three steps of the installation followed by a table to identify the changes corresponding to each component except UI. The instructions are as follows:

1. Download from the *gLite* web site the latest version of the CE installation script `xxxx\_installer.sh`. It is recommended to download the script in a clean directory.

2. Make the script executable (`chmodu+xxxxx\_install.sh`) and execute it.

3. Run the script as root. All the required RPMS are downloaded from the *gLite* software repository into the directory `xxxx` next to the installation script and the installation procedure is started. If one RPM is already installed, it is upgraded if necessary. Check the screen output for errors or warnings.

In Table C.1 the filenames that replace "xxxx" during the installation of each component in *gLite* are shown.

Component	Variable xxxx
Security	glite-security-utils
Computing Element	glite-ce
Worker Node	glite-wn
Workload Manager	glite-wms
Logging and Bookkeeping Server	glite-lb
gLite I/O server	glite-io-server
gLite I/O client	glite-io-client
Local Transfer Service	glite-data-local-transfer-service
Single Catalog	glite-data-single-catalog
R-GMA server	glite-rgma-server
R-GMA client	glite-rgma-client
R-GMA service-tool	glite-rgma-service-tool
R-GMA GadgetIN	glite-rgma-gin
VOMS server and administration tools	glite-voms-server

Table C.1: Filename variables during the installation of the different components of gLite.

C.2 Internal Workings of gLite

C.2.1 Information and Monitoring System

The R-GMA structure is divided into two main components: R-GMA server and R-GMA client. The R-GMA server comprises four components, namely the R-GMA server, the R-GMA schema server, the R-GMA registry server and the R-GMA browser. The R-GMA client comprises five components: the generic client, the R-GMA service-tool, the R-GMA site publisher, the R-GMA GadgetIn (GIN) and the R-GMA data archiver (flexible archiver). During the installation process of the R-GMA module, it is broken into four parts. The modules of Table C.2 are installed in sequence in order to fully deploy the whole R-GMAindexRelational grid monitoring architecture (R-GMA) module in gLite. Moreover, some rules have to be followed during the installation.

1. There must be exactly one schema servers for the grid.

Deployement component	Contains	Used/ included by
R-GMA server	R-GMA server R-GMA registry server R-GMA schema server R-GMA browser R-GMA site publisher R-GMA data archiver R-GMA service-tool	
R-GMA client	RGMA client APIs	User Interface(UI) Worker Node (WN)
R-GMA service-tool	R-GMA service-tool	Computing Element (CE) Data Local Transfer Service Data Single Catalog (MySQL) Data Single Catalog (Oracle) I/O-Server Logging and Bookkeeping (LB) R-GMA server Torque Server VOMS Server Workload Management System (WMS)
R-GMA GIN	R-GMA GadgetIN	Computing Element (CE)

Table C.2: Interaction between R-GMA and other grid resources.

2. There must be one or several registry servers per grid.

3. There must be one site publisher per site.

4. You can choose to enable/disable the data archiver.

After these steps, administrators can begin the installation for other services.

R-GMA Server

In the next few sections, we illustrate the procedures for the installation of the security mechanisms. The installation and the configuration of the R-GMA Server are also shown.

Security settings of R-GMA Server

1. Install one or more certificate authority certificates in /etc/ grid-security/certificates. The complete list of CA certificates can be downloaded in RPMS format from the Grid Policy Management Authority web site http://www.gridpma.org/.

2. Install the server host certificate hostcert.pem and key hostkey.pem in /etc/grid-security.

We now proceed with the installation procedures.

Installation of R-GMA Server

1. Follow the steps shown in Section C.1.2 in the paragraph entitled "During installation" for the initial steps of installation.

2. If the installation is performed successfully, the following components are installed:

```
gLite                   in opt/glite ($GLITE_LOCATION)
gLite-essentials-java   in $GLITE_LOCATION/externals/share
MySQL-server            in /usr
MySQL-client            in /usr
MySQL-shared-compat     in /usr
Tomcat                  in /var/lib/tomcat5
```

3. The *gLite* R-GMA server configuration script is installed in \$GLITE\ _LOCATION/etc/config/scripts/glite-rgma-server-config.py All the necessary template configuration files are installed into \$GLITE\ _LOCATION/etc/config/templates/.

Lastly, with the installation in place, the R-GMA server can be configured.

Configuration of R-GMA Server

1. Copy the global configuration file template `\$GLITE\_LOCATION/`
`etc/config/template/glite-global.cfg.xml` to `\$GLITE\_LOCATION/etc/`
`config`. Open it and modify the parameters if required.

2. Copy the R-GMA common configuration file template `\$GLITE\`
`_LOCATION/etc/config/templates/glite-rgma-common.cfg.xml` to `\$GLITE\`
`_LOCATION/etc/config` and modify the parameters values as necessary.
Some parameters have default values; others must be changed by the
user. All parameters that must be changed have a token value of
changeme.

3. Copy the R-GMA server configuration file template `\$GLITE\_LOCATION/`
`etc/config/templates/glite-rgma-server.cfg.xml` to `\$GLITE\_LOCATION/`
`etc/config` and modify the parameters values as necessary. Some parameters have default values; others must be changed by the user. All
parameters that must be changed have a token value of changeme.

4. As root run the R-GMA server configuration file `\$GLITE\_LOCATION/`
`etc/config/scripts/glite-rgma-server-config.py`.

R-GMA Client

In the next few sections, we illustrate the steps for setting up the security
mechanisms. The installation and the configuration of the R-GMA client are
also demonstrated.

Security Settings of R-GMA Client

Install one or more certificate authority certificates in `/etc/grid-security/`
`certificates`. The complete list of CA certificates can be downloaded in
RPMS format from the Grid Policy Management Authority web site `http:`
`//www.gridpma.org/`.

Next, we go on with the installation procedures.

Installation of R-GMA Client

1. Follow the steps shown in Section C.1.2 in the paragraph entitled "During installation" for the initial steps of installation.

2. If the installation is performed successfully, the following components
are installed:

```
gLite                    in /opt/glite( \$GLITE_LOCATION)
gLite-essentials-java    in \$GLITE_LOCATION/externals/share
gLite-essentials-cpp     in \$GLITE_LOCATION/externals/
swig-runtime             in \$GLITE_LOCATION/externals/
```

3. The *gLite* R-GMA configuration script is installed in `\$GLITE\` `_LOCATION/etc/config/scripts/glite-rgma-client-config.py`. All the necessary template configuration files are installed into `\$GLITE\` `_LOCATION/etc/config/templates/`.

Lastly, the R-GMA Client is configured.

Configuration of R-GMA Client

1. Copy the global configuration file template `\$GLITE\_LOCATION/` `etc/config/template/glite-global.cfg.xml` to `\$GLITE\_LOCATION/etc/` `config`, open it, and modify the parameters if required.

2. Copy the R-GMA common configuration file template `\$GLITE\` `_LOCATION/etc/config/templates/glite-rgma-common.cfg.xml` to `\$GLITE\` `_LOCATION/etc/config/glite-rgma-common.cfg.xml`, open it, and modify the parameter values as necessary. Some parameters have default values; others must be changed by the user. All parameters that must be changed have a token value of changeme.

3. Copy the R-GMA client configuration file template from `\$GLITE\` `_LOCATION/etc/config/templates/glite-rgma-client.cfg.xml` to `\$GLITE\` `_LOCATION/etc/config/glite-rgma-client.cfg.xml` and modify the parameter values as necessary. Some parameters have default values; others must be changed by the user. All parameters that must be changed have a token value of changeme. If you use the R-GMA client as a sub-deployment module that is downloaded and used by another deployment module, the configuration script is run automatically by the configuration script of the other deployment module and you can skip the following step. Otherwise continue.

4. Run the R-GMA Client configuration file `\$GLITE\_LOCATION/etc/` `config/scripts/glite-rgma-client-config.py`.

R-GMA Service-tool

In the following sections, we demonstrate the steps for setting up the security mechanisms. We then go on with the installation and the configuration of the R-GMA Service-tool.

Security Settings of R-GMA Service-tool

Install one or more certificate authority certificates in `/etc/grid-security/` `certificates`. The complete list of CA certificates can be downloaded in RPMS format from the Grid Policy Management Authority web site `http:` `//www.gridpma.org/`.

Next, the R-GMA Service-tool is installed.

Installation of R-GMA Service-tool

1. Follow the steps shown in Section C.1.2 in the paragraph entitled "During installation" for the initial steps of installation.

2. If the installation is performed successfully, the following components are installed:

   ```
   gLite                  in /opt/glite($GLITE_LOCATION)
   gLite-essentials-java  in $GLITE_LOCATION/externals/share
   ```

3. The *gLite* R-GMA service-tool configuration script is installed in `\$GLITE\_LOCATION/etc/config/scripts/glite-rgma-service-tool-config. py`. All the necessary template configuration files are installed into `\$GLITE\_LOCATION/etc/config/templates/`.

Finally, we configure the R-GMA Service-tool.

Configuration of R-GMA Service-tool

1. Copy the global configuration file template `\$GLITE\_LOCATION/etc/config/template/glite-global.cfg.xml` to `\$GLITE\_LOCATION/etc/config`, open it, and modify the parameters if required.

2. Copy the R-GMA common configuration file template from `\$GLITE\_LOCATION/etc/config/templates/glite-rgma-common.cfg.xml` to `\$GLITE\_LOCATION/etc/config/glite-rgma-common.cfg.xml`, open it, and modify the parameter values as necessary. Some parameters have default values; others must be changed by the user. All parameters that must be changed have a token value of changeme.

3. Copy the R-GMA service-tool configuration file template `\$GLITE\_LOCATION/etc/config/templates/glite-rgma-service-tool.cfg.xml` to `\$GLITE\_LOCATION/etc/config` and modify the parameter values as necessary. Some parameters have default values; others must be changed by the user. All parameters that must be changed have a token value of changeme.

4. Next you have to configure the information for each service that you want to be published by the R-GMA service-tool. The configuration parameters for the services are inside the configuration file of the service as separate instance parameter lists. In order to configure the service and the service-tool you have to modify the parameters for each of these "instance parameter lists".

5. You do not need to run the configuration script as this is done automatically by the configuration script of the deployment module that contains the corresponding services.

R-GMA GadgetIN (GIN)

In the following, we set up the security mechanisms for the R-GMA server. The R-GMA GadgetIN (GIN) is then installed and configured.

Security Settings

Install one or more certificate authority certificates in /etc/grid-security/ certificates. The complete list of CA certificates can be downloaded in RPMS format from the Grid Policy Management Authority web site http: //www.gridpma.org/.

Next, we install the R-GMA GadgetIN (GIN).

Installation of R-GMA GadgetIN

1. Follow the steps shown in Section C.1.2 in the paragraph entitled "During installation" for the initial steps of installation.

2. If the installation is performed successfully, the following components are installed:

   ```
   gLite                     in /opt/glite($GLITE_LOCATION)
   gLite-essentials-java     in $GLITE_LOCATION/externals/
       ...share
   ```

3. The *gLite* R-GMA gin configuration script is installed in \$GLITE\ _LOCATION/etc/config/scripts/glite-rgma-gin-config.py. All the necessary template configuration files are installed into \$GLITE_LOCATION/ etc/config/templates/.

Finally, the configuration procedures are carried out.

Configuration of R-GMA GadgetIN

1. Copy the global configuration file template \$GLITE_LOCATION/ etc/config/template/glite-global.cfg.xml to \$GLITE_LOCATION/etc/ config, open it, and modify the parameters if required.

2. Copy the R-GMA common configuration file template from \$GLITE\ _LOCATION/etc/config/templates/glite-rgma-common.cfg.xml to \$GLITE\ _LOCATION/etc/config/glite-rgma-common.cfg.xml, open it, and modify the parameter values as necessary. Some parameters have default values; others must be changed by the user. All parameters that must be changed have a token value of changeme.

3. Copy the GadgetIN configuration file template \$GLITE_LOCATION/etc/ config/templates/glite-rgma-gin.cfg.xml to \$GLITE_LOCATION/etc/ config and modify the parameter values as necessary. Some parameters

have default values; others must be changed by the user. All parameters that must be changed have a token value of changeme. If you use the R-GMA GadgetIN as a sub-deployment module that is downloaded and used by another deployment module (e.g., the CE), the configuration script is run automatically by the configuration script of the other deployment module and you can skip step 3. Otherwise continue.

4. Run the R-GMA GadgetIN configuration file `\$GLITE\_LOCATION/etc/config/scripts/glite-rgma-client-config.py`.

C.2.2 Workload Manager

The principal component in the WMS is the *Workload Manager*. It is involved in the submission and cancelation of jobs from the user interface. In the next few sections, we illustrate the procedures for the installation of the security mechanisms. The installation and the configuration of the workload manager are also shown.

Security Settings

1. Install one or more certificate authority certificates in `/etc/grid-security/certificates`. The complete list of CA certificates can be downloaded in RPMS format from the Grid Policy Management Authority web site `http://www.gridpma.org/`.

2. Customize the mkgridmap configuration file `\$GLITE\_LOCATION/etc/glite-mkgridmap.conf` by adding the required VOMS server group. The information in this file is used to run the `glite-mkgridmap` script during the security utilities configuration to produce the `/etc/grid-security/grid-mapfile`.

3. Install the server host certificate `hostcert.pem` and key `hostkey.pem` in `/etc/grid-security`.

We now proceed with the installation procedures.

Installation procedure

1. Follow the steps shown in Section C.1.2 in the paragraph entitled "During installation" for the initial steps of installation.

2. If the installation is performed successfully, the following components are installed:

```
gLite    in opt/glite
Condor   in /opt/condor-x.y.x (where x.y.z is the version)
Globus   in /opt/globus
```

3. The *gLite* wms configuration script is installed in \$GLITE_LOCATION/ etc/config/scripts/glite-wms-config.py. A template configuration file is installed in \$GLITE_LOCATION/etc/config/templates/glite-wms.cfg. xml.

4. The *gLite* WMS installs the R-GMA service-tool to publish its information to the information system R-GMA.

Lastly, with the installation in place, the Workload Manager can be configured.

Configuration

1. Copy the global configuration file template \$GLITE_LOCATION/ etc/config/template/glite-global.cfg.xml to \$GLITE_LOCATION/etc/ config, open it, and modify the parameters if required.

2. Copy the configuration file template from \$GLITE_LOCATION/etc/ config/templates/glite-wms.cfg.xml to \$GLITE_LOCATION/etc/config/ glite-wms.cfg.xml and modify the parameter values as necessary. Some parameters have default values; others must be changed by the user. All parameters that must be changed have a token value of changeme.

3. Configure the R-GMA service-tool. For this you have to configure the service-tool itself as well as configure the subservices of WMS for the publishing via the R-GMA service-tool:

   a. R-GMA service-tool configuration:
   Copy the R-GMA service-tool configuration file template \$GLITE\ _LOCATION/etc/config/templates/glite-rgma-service-tool.cfg.xml to \$GLITE_LOCATION/etc/config and modify the parameter values as necessary. Some parameters have default values; others must be changed by the user. All parameters that must be changed have a token value of changeme.

   b. Service Configuration for the R-GMA service-tool:
   Modify the R-GMA service-tool related configuration values that are located in the WMS configuration file glite-wms.cfg.xml that was mentioned before. In this file, you will find for each service that should be published via the R-GMA service-tool, one instance of a set of parameters that are grouped by the tag
   < instance name="xxxx" service="rgma-service-tool" >
   where xxxx is the name of corresponding subservice. For WMS the following subservices are published via the R-GMA service-tool and need to be updated accordingly. They are Locallogger, Proxy Renewal Service, Log Monitor Service, Job Controller Service, Network Server and *Workload Manager*.

4. Enter the VOMS server information for the VOs you want to use in the `glite-mkgridmap.conf` file in `/opt/glite/etc`.

5. As root run the WMS configuration file `/opt/glite/etc/config/scripts/glite-wms-config.py`.

C.2.3 Computing Element

The operations involved in the Computing Element are mainly that of the Job Management Service. It interacts essentially with the WMS or directly with the UI for operations on jobs. One significant operation is the broadcasting of resource information to the WMS for the matching process. This is done through the R-GMA mechanism.

Installation Prerequisites

In the following sections, we demonstrate the steps for setting up the security mechanisms in the Computing Element.

Security Settings

1. Install one or more certificate authority certificates in `/etc/grid-security/certificates`. The complete list of CA certificates can be downloaded in RPMS format from the grid policy management authority web site `http://www.gridpma.org/`.

2. Install the server host certificate `hostcert.pem` and key `hostkey.pem` in `/etc/grid-security`.

3. Install the VOMS Servers host certificate in the directory `/etc/grid-security/vomsdir`. This is necessary to allow LCMAPS to extract the VOMS information from the VOMS proxies.

4. The CE Service may require modification to the server firewall settings. The following iptables instructions must be executed.

```
-I <Chain_Name> 1 -m state --state NEW -m tcp -p
   tcp -- dport 2119 -j ACCEPT
-I <Chain_Name> 2 -m state --state NEW -m tcp -p
   tcp --dport 3878 -j ACCEPT
-I <Chain_Name> 3 -m state --state NEW -m tcp -p
   tcp --dport 3879 -j ACCEPT
-I <Chain_Name> 4 -m state --state NEW -m udp -p
   udp --dport 3879 -j ACCEPT
-I <Chain_Name> 5 -m state --state NEW -m tcp -p
   tcp --dport 3882 -j ACCEPT
-I <Chain_Name> 6 -m state --state NEW -m udp -p
   udp --dport 1020 -j ACCEPT
-I <Chain_Name> 7 -m state --state NEW -m udp -p
   udp --dport 1021 -j ACCEPT
```

```
-I <Chain_Name> 8 -m state --state NEW -m udp -p
udp --dport 1022 -j ACCEPT
-I <Chain_Name> 9 -m state --state NEW -m udp -p
udp --dport 1023 -j ACCEPT
-I <Chain_Name> 10 -m state --state NEW -m tcp -p
tcp --dport 32768:65535
```

Resource Management System

The resource management system has to be installed on the CE before configuration of the CE can be carried out. The resource management system can either be installed on dedicated computer clusters or on the same computer cluster as the CE. We now go on with the installation and the configuration of the Resource Management System.

Installation procedures

1. Follow the steps shown in Section C.1.2 in the paragraph entitled "During installation" for the initial steps of installation.

2. If the installation is performed successfully, the following components are installed:

   gLite   in /opt/glite ($GLITE_LOCATION)

   Condor   in/opt/condor-x.y.x (where x.y.z is the version)

   Globus   in /opt/globus ($GLOBUS_LOCATION)

   Tomcat   in /var/lib/tomcat5 (standard JPP location)

3. The *gLite* CE configuration script is installed in \$GLITE_LOCATION/ etc/config/scripts/glite-ce-config.py. A template configuration file is installed in \$GLITE_LOCATION/etc/config/templates/glite-ce.cfg. xml.

4. The *gLite* CE configuration script is installed in \$GLITE_LOCATION/ etc/config/scripts/glite-ce-config.py. A template configuration file is installed in \$GLITE_LOCATION/etc/config/templates/glite-ce.cfg. xml.

Finally, we configure the Resource Management System.

Configuration

1. Copy the global configuration file template \$GLITE_LOCATION/ etc/config/template/glite-global.cfg.xml to \$GLITE_LOCATION/etc/ config, open it, and modify the parameters if required.

2. Copy the configuration file template from `\$GLITE\_LOCATION/etc/config/templates/glite-ce.cfg.xml` to `\$GLITE\_LOCATION/etc/config/glite-ce.cfg.xml` and modify the parameter values as necessary. Some parameters have default values, others must be changed by the user. All parameters that must be changed have a token value of changeme.

3. Configure the R-GMA service-tool. For this you have to configure the service-tool itself as well as configure the subservices of CE for the publishing via the R-GMA service-tool:

   a. R-GMA service-tool configuration:
      Copy the R-GMA service-tool configuration file template `\$GLITE\_LOCATION/etc/config/templates/glite-rgma-service-tool.cfg.xml` to `\$GLITE\_LOCATION/etc/config` and modify the parameter values as necessary. Some parameters have default values; others must be changed by the user. All parameters that must be changed have a token value of changeme.

   b. Service Configuration for the R-GMA service-tool:
      Modify the configuration values that are located in the CE configuration file `glite-ce.cfg.xml` that was mentioned before. In this file, you will find for each service that should be published via the R-GMA service-tool, one instance of a set of parameters that are grouped by the tag
      < instance name="xxxx" service="rgma-service-tool" >
      Where xxxx is the name of corresponding subservice. For CE the following subservices are published via the R-GMA service-tool and need to be updated accordingly. They are namely locallogger, gatekeeper and CE monitor.

4. Install the VOMS servers host certificates in the directory `/etc/grid-security/vomsdir`.

5. Create the CE static information file `ce-static.ldif` in the directory that you specified in the cemon.static parameter.

6. As root run the CE configuration file `/opt/glite/etc/config/scripts/glite-ce-config.py`.

Worker Node

A *Worker Node* is the basic computing element in the grid. It is composed of the *gLite* I/O service, the logging and book-keeping service, the R-GMA service and the WMS checkpointing library.

Installation Prerequisites

In the next few sections, we illustrate the prerequisites for installing the Worker Node. This includes steps for setting up the security mechanisms and the validation of the installation of the Resource Management System. The installation and the configuration of the Worker Node are also demonstrated.

Security Settings

Install one or more certificate authority certificates in /etc/grid-security/ certificates. The complete list of CA certificates can be downloaded in RPMS format from the grid policy management authority web site http://www.gridpma.org/.

Resource Management System

The *Resource Management System* has to be installed on the worker node before configuration of the worker node can be carried out.

Next, we go on with the installation procedures.

Installation procedures

1. Follow the steps shown in Section C.1.2 in the paragraph entitled "During installation" for the initial steps of installation.

2. If the installation is performed successfully, the following components are installed:

gLite I/0	in /opt/glite
gLite LB	in /opt/glite
gLite R-GMA	in /opt/glite
gLite WMS Checkpointing	in /opt/glite
Globus	in /opt/globus

3. The *gLite* WN configuration scripts are installed in \$GLITE_LOCATION/ etc/config/scripts/glite-wn-config.py. A template configuration file is installed in \$GLITE_LOCATION/etc/config/templates/glite-wn.cfg. xml. Since the WN is a collection of clients, the individual configuration scripts and files are also installed and they must be run.

Lastly, the Worker Node is configured.

Configuration

1. Copy the global configuration file template \$GLITE_LOCATION/ etc/config/template/glite-global.cfg.xml to \$GLITE_LOCATION/etc/ config, open it, and modify the parameters if required.

2. Copy the configuration file template from \$GLITE_LOCATION/etc/ config/templates/glite-wn.cfg.xml to \$GLITE_LOCATION/etc/config/ glite-wn.cfg.xml and modify the parameter values as necessary.

3. Some parameters have default values; others must be changed by the user. All parameters that must be changed have a token value of changeme.

4. Run the WN configuration file `\$GLITE\_LOCATION/etc/config/scripts/ glite-wn-config.py`.

C.2.4   Data Management

Local Transfer Service

The Local Transfer Service is responsible for the transfer of files between different grid storage elements. It is composed of several services, namely File Transfer Service and File Placement Service and a number of agents, namely the Checker, the Fetcher and the Data Integrity Validator. In the following, we set up the security mechanisms for the Local Transfer Service. The Local Transfer Service is then installed and configured.

Security Settings

1. Install one or more certificate authority certificates in `/etc/ grid-security/certificates`. The complete list of CA certificates can be downloaded in RPMS format from the Grid Policy Management Authority web site `http://www.gridpma.org/`.

2. nstall the server host certificate `hostcert.pem` and key `hostkey.pem` in `/etc/grid-security`.

Next, we install the Local Transfer Service .

Installation Procedure

1. Follow the steps shown in Section C.1.2 in the paragraph entitled "During installation" for the initial steps of installation.

2. If the installation is performed successfully, the following components are installed:

   ```
   gLite                      in/opt/glite
   Tomcat                     in /var/lib/tomcat5
   MySQL                      in /usr/bin/mysql
   ```

3. The *gLite* LTS configuration script is installed in `\$GLITE\_LOCATION/ etc/config/scripts/glite-data-local-transfer-service-config.py`. A template configuration file is installed in `\$GLITE\_LOCATION/etc/config/ templates/glite-data-local-transfer-service.cfg.xml`.

Finally, the configuration procedures are carried out.

Configuration

1. Copy the global configuration file template \$GLITE_LOCATION/
   etc/config/template/glite-global.cfg.xml to \$GLITE_LOCATION/etc/
   config, open it, and modify the parameters if required.

2. Copy the configuration file template from \$GLITE_LOCATION/etc/
   config/templates/glite-data-local-transfer-service.cfg.xml   to
   \$GLITE_LOCATION/etc/config/glite-data-local-transfer-service.
   cfg.xml and modify the parameter values as necessary.

3. Some parameters have default values; others must be changed by the
   user. All parameters that must be changed have a token value of
   changeme.

4. As root run the LTS configuration file \$GLITE_LOCATION/etc/config/
   scripts/glite-data-local-transfer-service-config.py.

Single Catalog

The Single Catalog provides the mappings between LFN, GUID and SURL.
There are currently 2 modules in *gLite* that work with different databases.
For the MySQl version, the catalog glite-data-single-catalog is used while
in Oracle version, the catalog glite-data-single-catalog-oracle is used. In
the next few sections, we illustrate the procedures for the installation of the
security mechanisms. The installation and the configuration of the Single
Catalog are also shown.

Security Settings

1. Install one or more certificate authority certificates in /etc/
   grid-security/certificates. The complete list of CA certificates can
   be downloaded in RPMS format from the Grid Policy Management Authority web site http://www.gridpma.org/.

2. Install the server host certificate hostcert.pem and key hostkey.pem in
   /etc/grid-security.

Oracle JDBC Drivers

For the Oracle version, JDBC drivers have to be installed before installation of Single Catalog can begin. Download them from the Oracle
website http://www.oracle.com/technology/software/tech/java/sqlj_jdbc/
htdocs/jdbc101040.html and install them under the directory \${CATALINA\
_HOME}/common/lib.

We now proceed with the installation procedures.

Installation Procedure

1. Follow the steps shown in Section C.1.2 in the paragraph entitled "During installation" for the initial steps of installation.

2. If the installation is performed successfully, the following components are installed:

   `gLite`   in `/opt/glite`

3. The *gLite* SC configuration script is installed in `\$GLITE\_LOCATION/etc/config/scripts/glite-data-single-catalog-config.py`. A template configuration file is installed in `\$GLITE\_LOCATION/etc/config/templates/glite-data-single-catalog.cfg.xml`.

Last, with the installation in place, the Single Catalog can be configured.

Configuration

1. Copy the global configuration file template `\$GLITE\_LOCATION/etc/config/template/glite-global.cfg.xml` to `\$GLITE\_LOCATION/etc/config`, open it, and modify the parameters if required.

2. Copy the configuration file template from `\$GLITE\_LOCATION/etc/config/templates/glite-data-single-catalog.cfg.xml` to `\$GLITE\_LOCATION/etc/config/glite-data-single-catalog.cfg.xml` and modify the parameter values as necessary. Some parameters have default values; others must be changed by the user. All parameters that must be changed have a token value of changeme.

3. As root run the Single Catalog configuration file `\$GLITE\_LOCATION/etc/config/scripts/glite-data-single-catalog-config.py`.

C.3   Logging and Book-Keeping Server

The Logging and Book-keeping Server allows for the tracking of major events in a job (job submission, matchmaking, execution of jobs, etc.). In the next few sections, we illustrate the steps for setting up the security mechanisms. The installation and the configuration of the Logging and Book-keeping Server are also demonstrated.

Security Settings

1. Install one or more certificate authority certificates in /etc/ grid-security/certificates. The complete list of CA certificates can be downloaded in RPMS format from the Grid Policy Management Authority web site http://www.gridpma.org/.

2. Install the server host certificate hostcert.pem and key hostkey.pem in /etc/grid-security.

Next, we go on with the installation procedures.

Installation procedure

1. Follow the steps shown in Section C.1.2 in the paragraph entitled "During installation" for the initial steps of installation.

2. If the installation is performed successfully, the following components are installed:

gLite	in /opt/glite
Globus	in /opt/globus

3. The *gLite* LB configuration script is installed in \$GLITE_LOCATION/ etc/config/scripts/glite-lb-config.py. A template configuration file is installed in \$GLITE_LOCATION/etc/config/templates/glite-lb.cfg. xml.

4. The *gLite* LB installs the R-GMA service-tool to publish its information to the information system R-GMA.

Next, the Logging and Book-keeping Server is configured.

Configuration

1. Copy the global configuration file template \$GLITE_LOCATION/ etc/config/template/glite-global.cfg.xml to \$GLITE_LOCATION/etc/ config, open it, and modify the parameters if required.

2. Copy the configuration file template from \$GLITE_LOCATION/etc/ config/templates/glite-lb.cfg.xml to \$GLITE_LOCATION/etc/config/ glite-lb.cfg.xml and modify the parameter values as necessary. Some parameters have default values; others must be changed by the user. All parameters that must be changed have a token value of changeme.

3. Configure the R-GMA service-tool. For this you have to configure the service-tool itself as well as configure the subservices of LB for the publishing via the R-GMA service-tool:

   a. R-GMA service-tool configuration:

Copy the R-GMA service-tool configuration file template `\$GLITE\` `_LOCATION/etc/config/templates/glite-rgma-service-tool.cfg.` `xml` to `\$GLITE\_LOCATION/etc/config` and modify the parameters values as necessary. Some parameters have default values; others must be changed by the user. All parameters that must be changed have a token value of changeme.

   b. Service Configuration for the R-GMA service-tool:

Modify the R-GMA service-tool related configuration values that are located in the LB configuration file `glite-lb.cfg.xml` that was mentioned before. In this file, you will find for each service that should be published via the R-GMA service-tool, one instance of a set of parameters that are grouped by the tag

< instance name="xxxx" service="rgma-service-tool">

where xxxx is the name of the corresponding subservice. For LB the following subservices are published via the R-GMA service-tool and need to be updated accordingly. It is called the Log Server.

4. As root run the LB configuration file `\$GLITE\_LOCATION/etc/config/` `scripts/glite-lb-config.py`.

C.4 Security Mechanism

The security service carries out the authentication operations based on the certificate mechanism mentioned in the earlier chapters. In the following sections, we demonstrate the steps for the installation and the configuration of the Security Mechanism.

Installation Procedure

The installation of the *gLite* security module is carried out in the following procedure:

1. Follow the steps shown in Section C.1.2 in the paragraph entitled "During installation" for the initial steps of installation.

2. If the installation is performed successfully, the following components are installed:

```
CA Certificates        in /etc/grid-security/certificates
glite                  in /opt/glite ($GLITE_LOCATION)
```

Next, we configure the Security Mechanism.

Configuration

1. Copy the global configuration file template `\$GLITE\_LOCATION/etc/config/template/glite-global.cfg.xml` to `\$GLITE\_LOCATION/etc/config`, open it, and modify the parameters if required.

2. Copy the configuration file template `\$GLITE\_LOCATION/etc/config/template/glite-security-utils.cfg.xml` to `\$GLITE\_LOCATION/etc/config`, open it, and set all user-defined parameters. You can also modify, Advanced and System parameters if required, but it's normally not necessary. All parameters that must be changed have a token value of changeme.

3. Insert the appropriate entries in the `\$GLITE\_LOCATION/etc/glite-mkgridmap.conf` file, if grid-mapfile management is required.

4. Run the configuration script `\$GLITE\_LOCATION/etc/config/scripts/glite-security-utils-config.py`.

C.5 I/O

C.5.1 gLite I/O Server

In the following, we set up the security mechanisms for the gLite I/O Server. The gLite I/O Server is then installed and configured.

Security Settings

1. Install one or more certificate authority certificates in `/etc/grid-security/certificates`. The complete list of CA certificates can be downloaded in RPMS format from the Grid Policy Management Authority web site `http://www.gridpma.org/`.

2. Customize the `mkgridmap` configuration file `\$GLITE\_LOCATION/etc/glite-mkgridmap.conf` by adding the required VOMS server groups. The information in this file is used to run the `glite-mkgridmap` script during the Security Utilities configuration to produce the `/etc/grid-security/grid-mapfile`.

3. Install the server host certificate `hostcert.pem` and key `hostkey.pem` in `/etc/grid-security`.

Castor SRM

With some configuration of the Castor SRM, it is necessary to register the host DN of the *gLite* I/O Server in the Castor SRM server gridmap-file.

Next, we install the gLite I/O Server.

Installation Procedure

1. Follow the steps shown in Section C.1.2 in the paragraph entitled "During installation" for the initial steps of installation.

2. If the installation is performed successfully, the following components are installed:

   ```
   gLite Server          in /opt/glite
   Globus                in /opt/globus
   ```

3. The *gLite* I/O server configuration script is installed in `\$GLITE\_LOCATION/etc/config/scripts/glite-io-server-config.py`. A template configuration file is installed in `\$GLITE\_LOCATION/etc/config/templates/glite-io-server.cfg.xml`.

4. The *gLite* I/O server installs the R-GMA service-tool to publish its information to the information system R-GMA.

Finally, we configure the gLite I/O Server.

Configuration

1. Copy the global configuration file template `\$GLITE\_LOCATION/etc/config/template/glite-global.cfg.xml` to `\$GLITE\_LOCATION/etc/config`, open it, and modify the parameters if required.

2. Copy the configuration file template from `\$GLITE\_LOCATION/etc/config/templates/glite-io-server.cfg.xml` to `\$GLITE\_LOCATION/etc/config/glite-io-server.cfg.xml` and modify the parameter values as necessary. Some parameters have default values; others must be changed by the user. All parameters that must be changed have a token value of changeme.

3. Configure the R-GMA service-tool:
   Copy the R-GMA service-tool configuration file template `\$GLITE\_LOCATION/etc/config/templates/glite-rgma-service-tool.cfg.xml` to `\$GLITE\_LOCATION/etc/config` and modify the parameter values as necessary. Some parameters have default values; others must be changed by the user. All parameters that must be changed have a token value of changeme.

4. As root run the *gLite* I/O server configuration file `\$GLITE\_LOCATION/etc/config/scripts/glite-io-server-config.py`.

C.5.2 gLite I/O Client

This provides the APIs with the necessary interface to access resources using *gLite* I/O server. In the next few sections, we illustrate the procedures for the installation of the security mechanisms. The installation and the configuration of the gLite I/O Client are also shown.

Security Settings

Install one or more certificate authority certificates in `/etc/grid-security/certificates`. The complete list of CA certificates can be downloaded in RPMS format from the Grid Policy Management Authority web site `http://www.gridpma.org/`. We now proceed with the installation procedures.

Installation Procedure

1. Follow the steps shown in Section C.1.2 in the paragraph entitled "During installation" for the initial steps of installation.

2. If the installation is performed successfully, the following components are installed:

    ```
    gLite         in /opt/glite
    Globus        in /opt/globus
    ```

3. The *gLite* I/O client configuration script is installed in `\$GLITE\_LOCATION/etc/config/scripts/glite-io-client-config.py`. A template configuration file is installed in `\$GLITE\_LOCATION/etc/config/templates/glite-io-client.cfg.xml`.

Lastly, with the installation in place, the gLite I/O Client can be configured.

Configuration

1. Copy the global configuration file template `\$GLITE\_LOCATION/etc/config/template/glite-global.cfg.xml` to `\$GLITE\_LOCATION/etc/config`, open it, and modify the parameters if required.

2. Copy the configuration file template `\$GLITE\_LOCATION/etc/config/templates/glite-io-client.cfg.xml` to `\$GLITE\_LOCATION/etc/config/` and modify the parameter values as necessary. Some parameters have default values; others must be changed by the user. All parameters that must be changed have a token value of changeme.

3. As root run the *gLite* I/O client configuration file `\$GLITE\_LOCATION/etc/config/scripts/glite-io-client-config.py`.

C.5.3　User Interface

The user interface provides an interface for users to access services on the grid. For this to be possible, UI contains a number of components: *Data Catalog command-line clients* and APIs, *Data Transfer command-line clients* and APIs, gLite I/O client and APIs, R-GMA client and APIs, VOMS command-line tools, Workload Managemenet System clients and APIs and, Logging and Book-keeping clients and APIs. In the next few sections, we illustrate the steps for setting up the security mechanisms. The installation and the configuration of the user interface are also demonstrated.

Security Settings

The software concerning security settings can be downloaded from the web site `http://www.gridpma.org/`. Next, we continue with the installation procedures.

Installation Procedure

The UI can be installed in two ways. It can be installed as a root user or as a non-privileged user. The installation of these two types does not differ other than the directory in which the UI is installed.

1. Download from the *gLite* web site the latest version of the UI installation script `glite-ui\_install.sh`. It is recommended to download the script to a clean directory.

2. Make the script executable (`chmodu+xglite-ui\_installer.sh`) ! and execute it. If needed, pass the `Cbasedir<path>` option to specify the target installation directory.

3. Run the script as root or as normal user. All the required RPMS are downloaded from the *gLite* software repository in the directory `glite-ui` next to the installation script, and the installation procedure is started. If some RPM is already installed, it's upgraded if necessary. Check the screen output for errors or warnings. This step can fail if some of the OS RPMs are missing. If that is the case these RPMs must be installed manually from the OS distribution CD, or by apt/yum tools.

4. If the installation is performed successfully, the following components are installed:

   a. root installation:

   ```
   gLite          in /opt/glite(= GLITE_LOCATION)
   Globus         in /opt/globus(= GLOBUS_LOCATION)
   Gpt            in /opt/gpt(= GPT_LOCATION)
   ```

b. user installation:

gLite, Globus and GPT are installed in the tree from 'pwd'/ glite_ui by removing the /opt/[glite,globus,gpt] part. The GLITE_LOCATION, GLOBUS_LOCATION and GPT_LOCATION variables are set to the 'pwd'/glite_uivalue.

5. The *gLite* UI configuration script is installed in \$GLITE_LOCATION/ etc/config/scripts/glite-voms-server-config.py. A template configuration file is installed in \$GLITE_LOCATION/etc/config/templates/ glite-ui.cfg.xml.

Finally, the configuration procedures are carried out.

Configuration

1. Copy the global configuration file template \$GLITE_LOCATION/ etc/config/template/glite-global.cfg.xml to \$GLITE_LOCATION/etc/ config, open it, and modify the parameters if required.

2. Copy the configuration file templates from

```
$GLITE_LOCATION/etc/config/templates/glite-ui.cfg.xml to
$GLITE_LOCATION/etc/config/glite-ui-service.cfg.xml

$GLITE_LOCATION/etc/config/templates/glite-io-client.cfg.xml
... to
$GLITE_LOCATION/etc/config/glite-io-client.cfg.xml

$GLITE_LOCATION/etc/config/templates/glite-rgma-client.cfg.
...xml to
$GLITE_LOCATION/etc/config/glite-rgma-client.cfg.xml

$GLITE_LOCATION/etc/config/templates/glite-rgma-common.cfg.
...xml to
$GLITE_LOCATION/etc/config/glite-rgma-common.cfg.xml

$GLITE_LOCATION/etc/config/templates/glite-security-utils.
...cfg.xml to
$GLITE_LOCATION/etc/config/glite-security-utils.cfg.xml
```

and modify the parameter values as necessary for glite-io-client, glite-rgma-client, glite-rgma-common and glite-security-utils configuration files. Alternatively, a site configuration file can be used.

3. Some parameters have default values; others must be changed by the user. All parameters that must be changed have a token value of changeme. The configuration file contains one or more < set > sections, one per each VO that the UI must be configured for. It also contains a global < parameters > section.

4. Run the UI configuration file GLITE_LOCATION/etc/config/scripts/ glite-ui-config.py.

C.6   VOMS Server and Administration Tools

In the following sections, we demonstrate the steps for setting up the security mechanisms. We then go on with the installation and the configuration of the VOMS server and administration tools.

Security Settings

1. Install one or more certificate authority certificates in `/etc/grid-security/certificates`. The complete list of CA certificates can be downloaded in RPMS format from the Grid Policy Management Authority web site `http://www.gridpma.org/`.

2. Install the server host certificate `hostcert.pem` and key `hostkey.pem` in `/etc/grid-security`.

Next, the VOMS server and administration tools is installed.

Installation Procedure

1. Follow the steps shown in Section C.1.2 in the paragraph entitled "During installation" for the initial steps of installation.

2. If the installation is performed successfully, the following components are installed:

   ```
   gLite          in /opt/glite
   Tomcat         in /var/lib/tomcat5
   ```

3. The *gLite* VOMS Server and VOMS Admnistration configuration script is installed in `\$GLITE\_LOCATION/etc/config/scripts/glite-voms-server-config.py`. A template configuration file is installed in `\$GLITE\_LOCATION/etc/config/templates/glite-voms-server.cfg.xml`.

Finally, we configure the VOMS server and administration tools.

Configuration

1. Copy the global configuration file template `\$GLITE\_LOCATION/etc/config/template/glite-global.cfg.xml` to `\$GLITE\_LOCATION/etc/config`, open it, and modify the parameters if required.

2. Copy the configuration file template from `\$GLITE\_LOCATION/etc/config/templates/glite-voms-server.cfg.xml` to `\$GLITE\_LOCATION/etc/config/glite-voms-service.cfg.xml` and modify the parameter values as necessary.

3. Some parameters have default values; others must be changed by the user. All parameters that must be changed have a token value of changeme. Since multiple instances of the VOMS Server can be installed on the same node (one per VO), some of the parameters refer to individual instances. Each instance is contained in a separate name `<instance/>` tag. A default instance is already defined and can be directly configured. Additional instances can be added by simply copying and pasting the ¡instance/¿ section, assigning a name and changing the parameters values as desired.

4. As root run the VOMS Server configuration file `\$GLITE\_LOCATION/etc/` `config/scripts/glite-voms-server-config.py`.

References

[183] Stephen Burke, Simone Campana, Antonio Delgados Peris, Flavia Donno, Patricia Mendez Lorenzo, Roberto Santinelli, and Andrea Sciaba. *gLite 3: user guide*. Worldwide LHC computing Grid, January 2007.

[184] Xavier Jeannin. Vue d'ensemble de l'architecture logicielle. Web Published, October 2006. Available online at: http://www.urec.cnrs.fr/ IMG/pdf/JTR2006-architecture_gLite.pdf (accessed January 1st, 2009).

[185] EGEE. Running on the grid using gLite 1.3. Web Published, August 2005. Available online at: http://egee-jra1.web.cern.ch/egee-jra1/ Documentation/Tutorial/EGEE-JRA1-TEC-584530-GLITETUTORIAL-v0-1. pdf (accessed January 1st, 2009).

[186] EGEE. R-GMA command line tool. Web Published, November 2005. Available online at: http://hepunx.rl.ac.uk/egee/jra1-uk/glite-r1.5/ command-line.pdf (accessed January 1st, 2009).

[187] Sergio Andreozzi. The gLite middleware. Web Published, February 2007. Available online at: http://omii-europe.forge.cnaf.infn.it/ _media/jra2/documentation/djra2.0/glite-training.ppt (accessed January 1st, 2009).

[188] EGEE. gLite data management. Web Published, January 2007. Available online at: http://www.apan.net/meetings/manila2007/ presentations/egee/gLite_datamanagement.ppt (accessed January 1st, 2009).

[189] EGEE. gLite installation guide. Web Published, 2007. Available online at: http://glite.web.cern.ch/glite/packages/R1.0/R20050331/ doc/installation_guide.html (accessed January 1st, 2009).

Glossary

A

Abstract job object (AJO)   is a Java object that allows users to define jobs independent from the system. The jobs created by the clients are encapsulated as AJO.

Aggregate directory   is a collective repository for the resources present in the grid. GRIS and GIIS are examples of aggregate directories in MDS.

Aggregator services   are the services built on top of the aggregator framework that use aggregator sources to collect data. They can be queried to find information about resources in the grid using XPath queries.

Aggregator sources   are Java classes that are part of the aggregator framework of MDS. They implement an interface to collect XML formatted data from registered information providers (IPs).

AHEFT   is a heterogeneous earliest finish time-based adaptive rescheduling strategy for grid workflows. It reschedules the jobs in the workflow by monitoring the performance of the jobs in the workflow and by discovering newly available resources in the grid.

American options   are similar to European options but the owner of American options has the right to exercise the option at any time between the negotiation date t and the expiry date T.

Authentication   is the process by which an entity establishes its identity to the other entities in the network.

Authentication server (AS)   is a part of the Kerberos Key Distribution Centre (KDC), which shares a long-term secret key with all the entities in the Kerberos realm. It issues TGT to clients, which is used by them to obtain a ticket from the Ticket Granting Server for their actual communication with the desired server.

Authorization   is the verification of the privileges assigned to an entity to access the resources and services provided by other entities in the grid.

B

Backfill algorithm   is a scheduling algorithm that tries to find a job that can be started with the current available resources if the job at the head of the queue cannot be started due to lack of resource availability. This should not delay the scheduled start of the job at the head of the queue.

Berkeley database information index (BDII)   is an LDAP server that gathers information from individual GIIS. It contains a grid-level view of all the resources available in the grid.

Black and Scholes equation   is a partial differential equation in finance that governs the evolution of options prices under the assumption that the underlying stocks follow a stochastic process.

C

Certificate revocation list (CRL)   is a list of the serial numbers of the X.509 digital certificates that cannot be trusted because their validity has ended or because of some fraud.

Certifying authority (CA)   is a trusted third party in the PKI that issues digital certificates to individuals and organizations.

Chain of trust   is the process of trust establishment between an entity and a CA. This is done by verifying the correctness of the public key of the CA by tracing it upwards to another CA in the PKI hierarchy trusted by the entity.

Communication overhead   is the additional processing time spent by the system for control checking and error checking. For parallel computation, data exchange between the independent nodes constitutes the communication overhead.

Computing element (CE)   is a grid resource that carries out the execution of a job.

Condor pool   is a collection of agents, resources, and matchmakers.

Confidentiality refers to the hiding of sensitive information from the entities that do not have the rights to access them. It can be done either at the message level or at the transport level.

Credential delegation is the process of delegating one's complete or partial privileges to another entity in the grid. This allows the entity to access the resources on the behalf of the entity delegating the credential. Proxy certificates are used for credential delegation.

Cross realm authentication refers to the authentication of a security principal in one realm with a security principal in another realm. This is done by the sharing of a secret key between KDCs in the different Kerberos realm.

D

Data replication service (DRS) is a service to provide a pull-based replication capability for grid files. It is a high-level data management service built on top of two GT data management components: the Reliable File Transfer (RFT) Service and the Replica Location Service (RLS).

Directed acyclic graph (DAG) is a graphical representation of dependencies among tasks in a grid workflow. The nodes of a DAG represent the tasks and the directed edges represent the data dependencies.

Directed acyclic graph manager (DAGMan) is a meta-scheduler for the execution of programs (computations) in Condor.

Directory information tree (DIT) is a tree-based structure to organize names of entities in MDS in a hierarchical fashion.

Distributed fault-tolerant scheduling (DFTS) is a fault-tolerance mechanism for grids based on job replication.

E

European options are financial derivatives that give the owner of the option the right to buy (call) or sell (put) an underlying stock at a strike price K on the expiry date T negotiated at time t.

Exercise an option is the buying or selling of the underlying stock of an option at the strike price K.

Expected completion time is the wall-clock time at which a machine completes the execution of a task.

Expected execution time (EET) is the estimated time for the execution of a task on a machine when the machine has no job to execute.

Expected time to compute (ETC) matrix contains the expected execution time of tasks on all the machines in the grid. It is used by the scheduler to make mapping decisions.

Extensible markup language (XML) is a markup language whose purpose is to facilitate sharing of data across different interfaces using a common format.

External data representation (XDR) is a standard for the description and encoding of data. It is used to transfer data between different computer architectures.

F

Fast greedy is a mapping heuristic that assigns tasks to machines in an arbitrary order having the minimum completion time for that task. It is also known as MCT heuristic.

File transfer service (FTS) is a grid component that facilitates the transfer of data between different storage elements in the grid.

G

Genetic algorithm heuristic is a mapping heuristic that uses a genetic algorithm to find the best schedule for a metatask.

Globus access to secondary storage (GASS) allows access to data stored in a remote filesystem. Its client libraries allow applications to access remote files and its server component allows any computer to act as a limited file server.

Globus gridmap file contains a list of global names of users who are authorized to access a service on that node. All authorized users are mapped to a local user.

Globus resource allocation manager (GRAM) processes the requests for resources for remote application execution, allocates the required resources, and manages the active jobs. It also returns updated information regarding the capabilities and availability of the computing resources to the Metacomputing Directory Service (MDS).

Grid fabric is a layer containing the grid resources such as computional power, data storage, sensors, and network resources.

Grid index information service (GIIS) is a higher-level aggregate directory that collects information about grid resources from GRIS and lower-level GIIS. It contains the grid-level view of the resources.

Grid information protocol (GRIP) is a protocol for the discovery of new grid resources and enquiry of known grid resources. MDS uses LDAP for GRIP.

Grid monitoring architecture (GMA) is a consumer-producer based architecture for information and monitoring services in grids. It contains a registry for storing information about consumers and producers. It forms the basis of R-GMA.

Grid portal is an interface to a grid system. Users interact with the portal using an intuitive interface through which they can view files, submit and monitor jobs, and view accounting information.

Grid registration protocol (GRRP) is used by various components of MDS such as IPs and GRIS to inform other components about its existence.

Grid resource information service (GRIS) is a lower-level aggregate directory that collects information about grid resources from the information providers.

Grid resources are the components of a grid that are used in the processing of a job, e.g., computing element, storage element, etc.

Grid security infrastructure file transfer protocol (GSIFTP) is a file transfer protocol built on top of the grid security infrastructure. It allows secure data transfer between grid components.

Grid security infrastructure (GSI) is a part of the Globus Toolkit and defines the necessary standards for the implementation of security in grids. It consists of X.509 digital certificates, SAML and Globus gridmap file, X.509 proxy certificates and message protection using TLS or WS-security and WS-secure conversation.

Grid service is a stateful web service to make it suitable for grid applications.

Grid service handle (GSH) is used to distinguish different grid service instances of the same service created by the factory.

Grid service reference (GSR) contains grid service instance-specific information such as protocol binding, method definition and network address.

Grid workflow is the automation of a collection of services (or tasks) by coordination through control and data dependencies to generate a new service.

Grid workflow management system (GWFMS) is software used for modeling tasks and their dependencies in a workflow and managing the execution of workflow and their interaction with other grid resources.

GridFtp is a protocol defined by global grid forum-based on FTP. It provides secure, robust, fast and efficient transfer of bulk data in the grid environment. The Globus Toolkit provides the most commonly used implementation of the protocol.

GridRPC is a programming model based on client-server remote procedure call (RPC).

H

Heterogeneous computing (HC) refers to a system in which diverse resources are combined together to increase the combined performance and cost-effectiveness of these resources. Grid is an example of an HC system.

High-level Petri-net (HLPN) provides an extension to the classical Petri-nets by adding support to model data, time and hierarchy. It allows computation of output tokens of a transition based on multiple input tokens contrary to classical Petri-nets, which allow only one type of token.

High performance storage system (HPSS) is software that manages petabytes of data on disks and robotic tape libraries.

I

Information provider (IP) is a service that interfaces a data collection service and provides information about the available resources to the aggregate directories.

Information service is one of the main components of the grid. It provides static and dynamic information about the grid resources.

Information supermarket is a component in the workload manager that stores information about all active resources in the grid, which are used by the matchmaker for the decision-making process.

Interlogger helps to propagate the logging or book-keeping information from a grid component to the central book-keeping server.

J

Job is a computational task that is executed on the grid. The information pertaining to a job is specified by the user using a job description language.

Job description language (JDL) is a computer language used for describing a job based on information specified by the user.

Job handler is a component in the workload manager that carries out the packaging, submission, cancelation and monitoring of a job.

Job replication is a fault-tolerance strategy in which more than one copy of the same job are assigned to a different set of resources. The different instances may either run the same copy of the job or a copy using a different algorithm for the same job.

Job submission description language (JSDL) is a language to describe the requirements of a job for submission to grid resources. The job requirement is specified using XML.

K

Kerberos is a distributed authentication protocol that provides mutual authentication to client and server using symmetric-key cryptography.

Kerberos realm is the administrative domain in which Kerberos operates.

Key distribution center (KDC) is a trusted third party in Kerberos, which maintains a database containing account information for all the security principals in its realm. It consists of two components, Authentication Server and Ticket Granting Server.

L

LAPACK is a library of Fortran 77 subroutines for solving the most commonly occurring problems in numerical linear algebra. It has been designed to be efficient on a wide range of modern high-performance computers. The name LAPACK is an acronym for linear algebra package. The C version of LAPACK is SCLA-PACK.

LDAP is a protocol for querying and modifying directory services running over TCP/IP. LDAP support is implemented in web browsers and e-mail programs, which can query an LDAP-compliant directory. LDAP is a sibling protocol to HTTP and FTP and uses the `ldap://` prefix in its URL.

Level-based scheduling algorithm   is a scheduling algorithm, which partitions the DAG into levels of independent nodes and then uses heuristics like min-min, max-min and sufferage to map these nodes to processors.

List scheduling algorithm   is a scheduling algorithm, which assigns priorities to nodes in a DAG and considers the nodes with higher priority for scheduling before the lower priority nodes.

Local files catalog (LFC)   is a grid component that stores the mapping between different identifiers of a file or a resource.

LSF   is software for managing and accelerating batch workload processing for computationally intensive and data-intensive applications.

M

Machine availability time (MAT)   is the earliest time when a machine has completed the execution of all the previously assigned tasks and is ready to serve the next request.

Makespan   is defined as the maximum time taken for the completion of all the tasks in the metatask or for the execution of the complete grid workflow.

Mapper   is the component of a grid scheduler, which runs the mapping algorithm.

Mapping   is the overall process of matching and scheduling.

Masterworker   is a model for the execution of parallel applications in which a node (controlling master) sends pieces of work to other nodes (workers). The worker node performs the computation and sends the result back to the master node. A piece of work is assigned to the first worker node that becomes available next.

Matching   is the process of identifying suitable machines for a task.

Matchmaker   is a component that performs the matching of a job to grid resources based on the user information on the job and the available information on grid resources.

Max-min   is a mapping heuristic that finds the minimum expected completion time for each task in a metatask and then assigns the task having the maximum expected completion time to the corresponding machine.

MCT   see fast greedy.

Message passing interface (MPI) is a library of subroutines for handling communication and synchronization of programs running on parallel platforms.

MET see user-directed assignment.

Metacomputing uses many networked computers together as a single computational unit to provide massive processing power.

Metatask is a collection of independent tasks mapped to a collection of resources during a mapping event.

Middleware is a collection of software and packages used for the implementation of a grid.

Min-min is a mapping heuristic that finds the minimum expected completion time for each task in a metatask and then assigns the task having the least expected completion time to the corresponding machine.

Mixed-machine system is a class of HC system, which consists of heterogeneous machines connected by high-speed networks.

Monitoring and discovery services (MDS) is a component of the Globus Toolkit, which provides resource monitoring and discovery services within the grid environment.

Monte Carlo method is a stochastic computational method used for the simulation of various physical nondeterministic systems.

Monte Carlo trajectory is one possible stochastic simulation of the system with a given set of parameters.

MyProxy is an open-source software used for managing user X.509 certificates. It can be used to store and retrieve user credentials over the network in a secure way.

N

Namespace is a naming context in which each name should be unique.

Navier-Stokes equation is a partial differential equation in fluid mechanics that governs the motion of a viscous fluid given the various parameters.

Network job supervisor (NJS) is one of the Unicore components. It translates the jobs represented as AJO into target system-specific batch jobs. It also passes sub-AJOs to peer systems, synchronizes the execution of dependent jobs and manages the data transfer between different systems.

Node is a portion of the grid where a job can be executed independently on the grid. The parallel structure of the grid comes from running in parallel jobs on different nodes simultaneously.

O

Open grid services architecture (OGSA) defines a web services-based framework for the implementation of grid.

Open grid services infrastructure (OGSI) is a formal and technical specification of the implementation of grid services as defined by the OGSA framework.

Opportunistic load balancing (OLB) is a mapping heuristic that assigns tasks to the next available machine.

P

Parallel virtual machine (PVM) is a software package that permits a heterogeneous collection of Unix and/or Windows computers hooked together by a network to be used as a single large parallel computer.

Parameter sweep application is an application that executes multiple instances of a program using different sets of parameters and then collects the results from all the instances. Such applications frequently occur in scientific and engineering problems.

Petri-net is a modeling language that graphically represents the state of workflow in grids or distributed systems using the concept of tokens. It consists of places, transitions and directed arcs connecting the places to the transitions.

Pluggable authentication module (PAM) is a mechanism to integrate low-level authentication schemes with a high-level API so that the application may be written independent of the underlying authentication scheme. For example a MyProxy server can be configured to use an external authentication like an LDAP server.

Portable batch system (PBS) is a batch job and computer system resource management package. It accepts batch jobs (shell scripts with control attributes) and stores the job until it is run. It runs the job and delivers the output back to the user.

Portlet is a pluggable user interface component that is managed and displayed in a web portal.

Principal is the entity whose identity is being verified.

Proxy certificate is a part of the GSI, which is used by an entity to delegate its complete or partial privileges to another entity. It is also used for single sign-on. It has the same format as an X.509 digital certificate.

Public key infrastructure (PKI) is a method of secure communication between two entities in the internet using the public/private key pair. It consists of a trusted third party called the Certifying Authority (CA).

R

Relational grid monitoring architecture (R-GMA) is an information and monitoring system for grids developed by the European Data-Grid project. It is based on GMA and derives its flexibility from the relational model.

Reliable file transfer (RFT) is a web service that provides interface for controlling and monitoring third party file transfers using GridFTP. The client controlling the transfer is hosted inside a grid service so that it can be managed using the soft state model and queried using the ServiceData interface available to all grid services.

Remote procedure call (RPC) is a protocol that allows a program running on one host to invoke a procedure on a different host in the network.

Replica location service (RLS) is a service that allows the registration and location of replicas in Globus. It maps the logical file name to a physical file name.

Rescheduling is the process of assigning a job to a new machine, either to improve its performance or for the purpose of fault tolerance.

Risk neutral is a situation where investors do not take into consideration the risk when deciding an investment strategy. Investors are just concerned about the expected return.

S

Scheduling is the process of ordering the execution of a collection of tasks on a pool of resources.

Secure socket layer (SSL) is a protocol for secure communication over the Internet. SSL uses a public/private key pair for the encryption and decryption of the data. The public key is known to everyone and the private or secret key is known only to the recipient of the encrypted message.

Security assertion markup language (SAML) is an XML-based standard protocol that supports the exchange of identity information under different environments. Identity information is exchanged as assertions between the provider and consumer of assertions.

Security principal is any entity in a Kerberos realm. It can be a client, a server or the KDC.

Service level agreement (SLA) defines the minimum quality of sevice, availability and other service-related attributes expected by the user from the service provider and the charges levied on them.

Simple object access protocol (SOAP) is an XML-based communication protocol, which can be used by two parties communicating over the internet.

Single program multiple data (SPMD) is a style of parallel programming where all the processors use the same program but process different data.

Single sign-on is the process of authenticating once to obtain proxy credentials, which can be used to access grid resources without needing further authentication for a certain period.

Storage element (SE) is a grid resource that stores the information required or generated by the computing element.

Strike price is the predetermined price of the underlying stock at expiry date. This price is predetermined at time 0, which is the time when the option is bought.

Sufferage heuristic is a mapping heuristic that maps tasks in the decreasing order of sufferage value.

Sufferage value is the difference between best and second-best minimum completion time for a task.

Switching algorithm is a mapping heuristic that tries to strike a balance between MCT and MET heuristics by switching between the heuristics based on the load of the system.

T

Task farming is a type of parallel application, where many independent jobs are executed on machines around the world. Only a small amount of data needs to be retrieved from each of these jobs.

Task-level fault tolerance achieves fault tolerance by either rescheduling the job or by using a job replication strategy without affecting the workflow.

Task queue is a component of the workload manager that holds various jobs for the eventual allocation by the matchmaker. Authentication of the user information in the job is also done in the task queue.

Testbed is an experimental platform including dedicated hardware, software resources and scientific instruments. It is used to test and analyze the tools and products. It usually supports real-time deployment and interaction.

Ticket is a piece of information used by clients to authenticate to a server in Kerberos. Two kinds of tickets are used in Kerberos: Ticket Granting Ticket (TGT) issued by the AS to the client and a normal ticket issued by the TGS to the client.

Ticket granting server (TGS) is part of the Kerberos KDC that verifies the TGT issued to clients by the AS and issues session tickets to them to communicate for a specific duration with the desired server.

Ticket granting ticket (TGT) is issued by the AS to a client. The TGS verifies the validity of the TGT before issuing actual communication ticket to the client.

Trust is a relationship between two entities that forms the basis for the subsequent authentication and authorization between the two entities.

Trusted third party (TTP) is an entity that provides for the authentication of two parties, both of which trust the third party. CA is an example of a TTP.

U

Universal description, discovery and integration (UDDI) is an XML-based registry used for finding a web service on the Internet.

User-directed assignment is a mapping heuristic that assigns tasks in an arbitrary order to the machine having the minimum execution time for that task.

V

Verifier is the entity that verifies the identity of the principal.

Virtual organization (VO)   is a dynamic collection of multiple organizations that provides coordinated resource sharing. A grid usually consists of multiple virtual organizations.

W

Web service definition language (WSDL)   is an XML document used to describe a web service interface.

Web service (WS)   is a software system designed to support interoperable machine-to-machine interaction over a network.

Workflow-level fault tolerance   allows changes to the workflow structure to achieve fault tolerance. These include user-defined exceptions and task crash failures that cannot be handled by the task-level failure handling techniques.

Workload manager (WM)   is an interface in gLite that deals with the allocation, collection and cancellation of a job. It also provides information about the job status and the grid resources.

WS-federation   is a specification for standardizing how organizations share user identities in a heterogeneous authentication and authorization system.

WS-policy   is a specification for the service requestor and service provider to enumerate their capabilities, needs and preferences in the form of policies.

WS-privacy   is a proposed web service specification. It will use a combination of WS-security, WS-policy and WS-trust for communicating privacy policies among organizations.

WS-resource framework (WSRF)   is a generic and open framework for modeling and accessing stateful resources using web services. It contains a set of six web services specifications that define what is termed as the WS-resource approach to model and manage stateful resources in a web service context.

WS-secure conversation   is a web service extension built on top of WS-security and WS-trust. It provides a security context for the protection of more than one related message.

WS-security   is the standard to provide security features such as integrity, privacy, confidentiality and single message authentication to SOAP messages.

WS-trust   is an extension to the WS-security specification. It defines additional constructs and primitives for the request and issue of security tokens. It also provides ways to establish trust relationships with parties in different trust domains.

X

X.509 digital certificate   is a standard for digital certificates described by the RFC 2459. It consists of the public key of the certificate owner and is signed by the certifying authority.

XML digital signature   is a way of digitally signing the SOAP messages to ensure their integrity.

XML encryption   is a standard that provides end-to-end security for applications requiring secure XML data exchange. A SOAP message body is encrypted using block encryption algorithms like AES-256.

XSufferage heuristic   is a modification of sufferage heuristics that takes into account the location of data, while making the scheduling decision. Instead of grid-level MCT, XSufferage heuristics use cluster-level MCT to find the sufferage value.

Index

T - #0379 - 071024 - C336 - 234/156/15 - PB - 9780367385828 - Gloss Lamination